Ecosystem Services in Agricultural and Urban Landscapes

Ecosystem Services in Agricultural and Urban Landscapes

Edited by

Steve Wratten

Bio-Protection Research Centre
Lincoln University, New Zealand

Harpinder Sandhu

School of the Environment
Flinders University, Australia

Ross Cullen

Department of Accounting, Economics and Finance
Lincoln University, New Zealand

Robert Costanza

Crawford School of Public Policy
Australian National University, Australia

WILEY-BLACKWELL

A John Wiley & Sons, Ltd., Publication

Library of Congress Cataloging-in-Publication Data

Ecosystem services in agricultural and urban landscapes / edited by Steve Wratten, Harpinder Sandhu, Ross Cullen, and Robert Costanza.
      pages   cm
   Includes bibliographical references and index.
   ISBN 978-1-4051-7008-6 (cloth)
1. Ecosystem services.   2. Ecosystem management.   3. Ecology–Economic aspects.   I. Wratten, Stephen D., editor of compilation.   II. Sandhu, Harpinder, editor of compilation.   III. Cullen, Ross, 1948– editor of compilation.   IV. Costanza, Robert, editor of compilation.
   QH541.15.E267E28 2012
   333.72–dc23
                                            2012033656

A catalogue record for this book is available from the British Library.

Wiley also publishes its books in a variety of electronic formats. Some content that appears in print may not be available in electronic books.

Main Cover image A baby spinach field in Dome Valley, Arizona. Courtesy of John C. Palumbo.

Inset images courtesy of Morguefile/Darren Hester, Clarita and Fractl.

Cover design by Steve Thompson

Set in 10.5/12pt Classical Garamond BT by SPi Publisher Services, Pondicherry, India
Printed and bound in Malaysia by Vivar Printing Sdn Bhd

1   2013

Contents

5 Aquaculture and Ecosystem Services: Reframing the Environmental and Social Debate 58
Corinne Baulcomb

6 Urban Landscapes and Ecosystem Services 83
Jürgen Breuste, Dagmar Haase and Thomas Elmqvist

Contributors

Onil Banerjee
CSIRO Ecosystem Sciences
PMB2
Glen Osmond
South Australia 5064
Australia

Ramesh Baskaran
Faculty of Commerce
PO Box 84
Lincoln University
Christchurch 7647
New Zealand

Corinne Baulcomb
Scottish Agricultural College
West Mains Road
Edinburgh EH9 3JG
Scotland

Jürgen Breuste
Department of Geography/Geology
University Salzburg
Hellbrunnerstrasse 34
A 5020 Salzburg
Austria

Anthea Coggan
CSIRO Ecosystem Sciences
GPO Box 2583
Brisbane 4102
Queensland
Australia

Robert Costanza
Crawford School of Public Policy
Crawford Building (132)
Australian National University
Canberra ACT 0200
Australia

Neville D. Crossman
CSIRO Ecosystem Sciences
PMB2
Glen Osmond
South Australia 5064
Australia

Ross Cullen
Department of Accounting,
Economics and Finance
PO Box 84
Lincoln University 7647
Christchurch
New Zealand

Rudolf S. de Groot
Environmental Systems Analysis
Group
Wageningen University
PO Box 47
6700 AA Wageningen
the Netherlands

Thomas Elmqvist
Department of Systems Ecology and
Stockholm Resilience Centre
Stockholm University
SE-106 91 Stockholm
Sweden

Dagmar Haase
Institute of Geography
Humboldt-University Berlin
Berlin
Germany
and
Helmholtz Centre for Environmental
Research GmbH-UFZ
Permoserstraße 15
04318 Leipzig
Germany

Mary Haropoulou
Faculty of Commerce
Lincoln University
PO Box 84
Christchurch 7647
New Zealand

Claus Holzapfel
Department of Biological Sciences
Rutgers University
Newark
New Jersey 07102
USA

Marco Jacometti
Bio-Protection Research Centre
PO Box 84
Lincoln University
Lincoln 7647
New Zealand

Nicholas Jordan
Agronomy and Plant Genetics
Department
University of Minnesota
411 Borlaug Hall
1991 Buford Circle
St. Paul
Minnesota 55018
USA

Pamela Kaval
Havelock North
New Zealand and
Marylhurst University
Oregon
USA

Sofia Orre-Gordon
Barbara Hardy Institute
University of South Australia
GPO Box 2471
Adelaide
South Australia 5001
Australia
and
Bio-Protection Research Centre
PO Box 84
Lincoln University
Lincoln 7647
New Zealand

John Porter
Department of Plant and
Environmental Science
Faculty of Life Sciences
University of Copenhagen (KU-LIFE)
HøjbakkegårdAlle 9
2630 Taastrup
Denmark

Jack Radford
Lincoln University
Faculty of Commerce
PO Box 84
Christchurch 7647
New Zealand

Harpinder Sandhu
School of the Environment
Flinders University
GPO Box 2100
Adelaide SA 5001
Australia

Clive Smallman
University of Western Sydney
School of Business
Locked Bag 1797
Penrith NSW 2751
Australia

Jean Tompkins
Bio-Protection Research Centre
PO Box 84
Lincoln University
Lincoln 7647
New Zealand

Keith Douglass Warner
Center for Science, Technology and
Society
Santa Clara University
500 El Camino Real
Santa Clara
California 95053
USA

Stuart M. Whitten
CSIRO Ecosystem Sciences
GPO Box 1700
Canberra
ACT 2601
Australia

Steve Wratten
Bio-Protection Research Centre
PO Box 84
Lincoln University
Lincoln 7647
New Zealand

Xiaoyong Yuan
Key Laboratory of Ecosystem
Network Observation and
Modelling
Institute of Geographic Sciences
and Natural Resources Research
Chinese Academy of Sciences
Beijing 100101
China

Yangjian Zhang
Key Laboratory of Ecosystem
Network Observation and
Modelling
Institute of Geographic Sciences
and Natural Resources
Research
Chinese Academy of Sciences
Beijing 100101
China

Reviewers

Editors acknowledge the contribution of following reviewers for their helpful comments and suggestions that helped to improve clarity of the chapters.

- Andrew Davidson, SEQ Catchments Ltd, Brisbane, Queensland, Australia.
- Brenda Lin, CSIRO Marine & Atmospheric Research, Melbourne, Victoria, Australia.
- Francis Turkelboom, Research Institute for Forest and Nature (INBO), Brussels, Belgium.
- Gupta Vadakattu, CSIRO Ecosystem Sciences, Adelaide, Australia.
- Uday Nidumolu, CSIRO Ecosystem Sciences, Adelaide, Australia.
- Yuki Takatsuka, Temple University, Japan.

Foreword

It is now becoming clear that an ecosystem approach is the most appropriate methodology to ensure sustainable food security and conservation of urban landscapes. Hence this book by Steve Wratten and colleagues is a timely one. At the time of the origin of agriculture or settled cultivation over 10 000 years ago, the early cultivators, mostly women, adopted an ecosystem approach for standardizing cultivation practices, as well as in the choice of crops. For example, in the state of Tamil Nadu in India, ancient scholars divided the state into five major agroecological zones. These were: coastal, hill, arid, semiarid and wet zones. Agricultural practices were followed according to the specific ecosystem, keeping in view the extent of rainfall, the incidence of sunlight and the moisture-holding capacity of the soil. From the naturally occurring biodiversity, plants with specialized adaptations, such as halophytes for coastal areas and xerophytes for the arid zone, were identified and cultivated.

An ecosystem approach to soil and water management helps to ensure successful agriculture. Water security is important not only for agriculture and industry, but also for domestic needs and for ecosystem maintenance. The book covers all aspects of soil health conservation and enhancement, and water and biodiversity management. Ecosystem-based agriculture ensures stability of production and at the same time enhances the coping capacity of farming families to meet the challenges of climate change. I therefore hope that this book will be widely read and used both by farming practitioners and policy makers. We owe a deep debt of gratitude to the editorial team for their dedication to the cause of sustainable agriculture and food security.

M. S. Swaminathan

PROF M S SWAMINATHAN
Member of Parliament (Rajya Sabha)
Emeritus Chairman, M S Swaminathan Research Foundation
Third Cross Street, Taramani Institutional Area
Chennai - 600 113 (India)

Introduction

Ecosystem goods and services provide mankind with most necessities of life and survival. They include such processes as biological control of pests, weeds and diseases, pollination of crops, amelioration of flooding and wind erosion, provision of food (including fisheries), the hydro-geochemical cycle, capture of carbon by plants and by soil and providing settings for much of the world's tourism. A pivotal paper by Robert Costanza and colleagues written in 1997 used a range of methods to quantify ecosystem services (ES) and to estimate their total economic value worldwide. The estimate was $US33 trillion ($10^{12}$) per annum. Costanza et al.'s valuation stimulated much debate, including the suggestion that $US33 trillion is 'a serious underestimate of infinity'. In other words, some people believe that mankind cannot survive without ES, so evaluating it is futile. However, ES world-wide are being degraded more rapidly than ever before and this degradation poses serious threats to quality of life and therefore to modern economies. The *Millennium Ecosystem Assessment* (MEA) pointed to the very high rate of ES loss and the consequences for global stability if that rate continues.

In the same year as Costanza et. al.'s paper, Gretchen Daily, of Stanford University, USA, published a key book entitled *Nature's Services*. Those two publications led to a change in the paradigm within which mankind's dependence on living things is viewed. However, Costanza and Daily concentrated largely on 'natural' ecosystems and biomes, such as boreal forests, coral reefs and mangroves. They did not concentrate on the many ecosystem services provided by highly modified or managed areas, such as farmland and cities. However, ES in these systems are of vital significance to the survival and productivity of those systems, as more than 50% of the world's population lives in cities and this proportion is increasing by 1–2% per annum. The 'ecological footprint' of cities is enormous and, with cities such as Shanghai forecast to grow from 17 million to 70 million over the next decade, the extent to which cities can support themselves in even a limited number of ecosystem functions is likely to continue to decline.

ES underpin life on earth, provide major inputs to many sectors of the economy and support our lifestyles. This book explores the role that ES play in two settings where humans have actively modified ecological systems: agriculture and urban areas. It addresses the hitherto under-estimation of ES in farmland and cities and explores ways to develop concepts, policies and methods of evaluating ES, as well as the ways in which ES in these systems can be maintained and enhanced. This approach is timely and will be of high scientific and political value, especially given that the MEA disappeared from world media and discussion very soon after it was announced, because of a widely-held but increasingly erroneous belief that technology will rescue mankind as the environmental equivalent of 'peak oil' is approached.

The book is divided into four parts with a series of self-contained chapters connected by the overall aim of the book. The Introduction is written by the editorial team to highlight the importance of ES in natural and managed landscapes. Part A sets the scene by introducing the concept of ES in managed landscapes such as farmland and cities. Chapter 1 explains the concept of ES and their importance. Chapter 2 provides links between ecosystem function to economic benefits by exploring changes in these due to change in land and water management. Chapter 3 deals with key concepts and methods to value ES. Part B provides information on ES in three different managed systems: viticulture, aquaculture and urban areas. Chapter 4 discusses ES associated with viticulture and techniques to enhance them. Chapter 5 explores environmental and social impacts of aquaculture and maps them through an ES typology. Chapter 6 develops the concept of ES in urban planning and management. It discusses ES relevant to urban areas and their importance in planning and management of cities. Part C focuses on measuring and monitoring ES at different scales. Chapter 7 develops this theme by also exploring ES at a range of spatial scales with case studies ranging from landscape, to regions and biomes. Chapter 8 provides frameworks to evaluate ES using 'bottom-up' field-scale measurements. It also discusses scenarios for balancing production and ES on farmland. Part D discusses design of ecological systems for the delivery of ES. In this Part, Chapter 9 explores the concept of multifunctional agriculture in the Upper Midwest region of the US. Chapter 10 discusses the role of ES through supply chain management in a wool enterprise. Chapter 11 analyses the concept of market-based instruments by providing examples to improve the delivery of ES. The epilogue examines prospects for the future and the role of ES in contributing to sustainable agriculture and cities.

We believe this book will be useful to senior undergraduates, postgraduates, environmental economists, agriculturalists, theoretical and applied ecologists, local and regional planners and government personnel in understanding the role of ES in a sustainable future. This book has been written by an international team of researchers. We acknowledge the effort, expert knowledge and care of team members that brought this project to completion and sincerely thank all of the authors for their contributions. The editors thank their family and friends for their continued support.

We end this Introduction with one of our favourite quotations about ES and 'future farming': 'I am a photosynthesis manager and an ecosystem-service provider', Peter Edlin, farmer, Sweden, 2003.

Steve Wratten (Lincoln), Harpinder Sandhu (Adelaide),
Ross Cullen (Lincoln), Robert Costanza (Canberra)
May 2012

Part A

Scene Setting

1

Ecosystem Services in Farmland and Cities

Harpinder Sandhu[1] and Steve Wratten[2]

[1]School of the Environment, Flinders University, Adelaide, Australia
[2]Bio-Protection Research Centre, Lincoln University, Lincoln, New Zealand

Abstract

Ecosystems sustain human life through the provision of four types of ecosystem services (ES) – a central tenet of the United Nations' Millennium Development Goals (MDGs). These categories are, with examples: supporting (water and nutrient cycling), provisioning (food production, fuel wood), regulating (water purification, erosion control), and cultural (aesthetic and spiritual values). A recent trend has been a decline in ES globally, largely due to ignorance of their value to human well-being and inadequate socioeconomic valuation mechanisms that encourage individuals/governments to invest in maintaining them. Engineered ecosystems from farmland and cities are the most important providers of ES for the world population. However, they are largely left outside the decision-making process in managing agriculture and urban areas, due to the general low awarencss of how the ES associated with these systems can and have been quantified. As nearly half of the world population is dependent on agriculture for its livelihood and cities are expanding at a faster rate than ever before, it is vital to understand, measure and incorporate ES into decision making and planning of agriculture and cities. This chapter discusses the concept of ES, their valuation methods, the types of engineered systems and how ES can be adopted by them to enhance them and ensure an equitable and sustainable future.

Ecosystem Services in Agricultural and Urban Landscapes, First Edition. Edited by Steve Wratten, Harpinder Sandhu, Ross Cullen and Robert Costanza.
© 2013 John Wiley & Sons, Ltd. Published 2013 by John Wiley & Sons, Ltd.

Introduction

Natural and modified ecosystems support human life through functions and processes known as ecosystem services (ES; Daily, 1997). These are the life-support systems of the planet (Myers, 1996; Daily, 1997; Daily et al., 1997) and it is evident that human life cannot exist without them.

The importance of ecosystem goods and services in supporting human life and as a life-support system of the planet (Myers, 1996; Daily, 1997; Costanza et al., 1997; Millennium Ecosystem Assessment, 2005) is now very well established and ES were demonstrated to be of very high economic value 15 years ago (US $33 trillion year^{-1}; Costanza et al., 1997). Although that value-transfer approach has been heavily criticized (Toman, 1998), no subsequent attempt to quantify ES globally has been made. However, for particular biological groups, such as insects, value transfer has again been used (Losey and Vaughan, 2006) or for one taxon for one region, experimental techniques to evaluate animals' populations have been combined with the economic value of the support they provide (e.g. earthworms and soil formation; Sandhu et al., 2008). Also, a whole-of-farm approach has been again based on in situ measurements followed by spatial scaling (Porter et al., 2009), in that case for the whole of the European Union in relation to current agricultural subsidies. Yet because most ES are not traded in economic markets, they carry no 'price tags' (no exchange value in spite of their high use value) that could alert society to changes in their supply or deterioration of underlying ecological systems that generate them. Despite this, there has been a recent trend of decline in ES globally, with 60% of the ES examined having been degraded in the last 50 years (Millennium Ecosystem Assessment, 2005). Global efforts to halt this decline in ES have increased considerably since the completion of the Millennium Ecosystem Assessment (MEA) in 2005. The United Nations has established the Intergovernmental Science-Policy Platform on Biodiversity and Ecosystem Services (IPBES) to translate science into action world-wide in consultation with governments and research partners (IPBES, 2010).

Because the threats to ES are increasing, there is a critical need for identification, monitoring and enhancement of ES both locally and globally, and for the incorporation of their value into decision-making processes (Daily et al., 1997; Millennium Ecosystem Assessment, 2005; IPBES, 2010; UN, 2012). It is well known that agroecosystems and urban areas contribute substantially to the welfare of human societies by providing highly demanded and valuable ES. Many of these, however, remain outside conventional markets. This is especially the case for public goods (climate regulation, soil erosion control, etc.) and external costs related to the active protection and management of these ecosystems. The capacity of ecosystems to deliver ES is already under stress (Millennium Ecosystem Assessment, 2005) and additional challenges imposed by climate change in the coming years will require better adaptation (Mooney et al., 2009).

What are ecosystem services?

The Millennium Ecosystem Assessment sponsored by the United Nations (Millennium Ecosystem Assessment, 2005) defines ecosystem services (ES) as the benefits people obtain from ecosystems. There is a general lack of understanding of what an ecosystem actually is, however; for example, among university undergraduates and even researchers it is probably worth remembering that single species can provide ES, albeit as part of their place in a trophic web. The facts that honey bees pollinate crops and ladybugs (ladybirds) eat insect pests are often a simple way of illustrating the power of ES to land owners, among others. In these circumstances, 'nature's services' can be a more useful phrase. These benefits sustain human existence through four types of service that include supporting (e.g. water and nutrient cycling), provisioning (e.g. food production, fuel wood), regulating (e.g. water purification, erosion control), and cultural (e.g. aesthetic and spiritual values) services. Benefits arise from managed as well as natural ecosystems. Recent studies have contributed to further understanding of ES for natural resource management (Wallace, 2007), for accounting purposes (Boyd and Banzhaf, 2007), for valuation (Fisher and Turner, 2008), and for policy-relevant research (Fisher et al., 2008; Balmford et al., 2011). Sagoff (2011) points out the differences in ecological and economic criteria in assessing and valuing ES and advocates for a conceptual framework to integrate market-based and science-based methods to manage ecosystems for human well-being.

Ecosystem functions, goods and services

Ecosystem functions can be defined as 'the capacity of natural processes and components to provide goods and services that satisfy human needs, directly or indirectly' (de Groot, 1992). Using this definition, ecosystem functions are best conceived as a subset of ecological processes and ecosystem structures. Each function is the result of the natural processes of the total ecological subsystem of which it is a part. Natural processes, in turn, are the result of complex interactions between biotic (living) and abiotic (chemical and physical) components of ecosystems through the universal driving forces of matter and energy (de Groot et al., 2002).

One of the key insights provided by the MEA (2005) is that not all ES are equal – there is no one single category that captures the diversity of what fully functioning ecological systems provide humans. Rather, researchers must recognize that ES occur at multiple scales, from climate regulation and carbon sequestration at the global scale, to soil formation and nutrient cycling more locally. To capture the diversity of ES, the MEA (2005) grouped them into four basic services based on their functional characteristics.

1 Regulating services: ecosystems regulate essential ecological processes and life support systems through biogeochemical cycles and other biospheric processes. These include climate regulation, disturbance moderation and waste treatment.
2 Provisioning services: the provisioning function of ecosystems supplies a large variety of ecosystem goods and other services for human consumption,

ranging from food in agricultural systems, raw materials and energy resources.

3 Cultural services: ecosystems provide an essential 'reference function' and contribute to the maintenance of human health and well-being by providing spiritual fulfilment, historical integrity, recreation sites and aesthetics.

4 Supporting services: ecosystems also provide a range of services that are necessary for the production of the other three service categories. These include nutrient cycling, soil formation and soil retention.

The ES framework

The ES framework has been increasingly used to explain the interactions between ecosystems and human well-being. Several studies classified ES into different categories based on their functions (Costanza et al., 1997; Daily, 1997; de Groot et al., 2002). The MEA assessed the consequences of ecosystem change for human well-being and provided a framework to identify and classify ES (Millennium Ecosystem Assessment, 2005). It established the scientific basis for actions needed to balance nature and human well-being by sustainable use of ecosystems. In the following section, we follow MEA typology and discuss the ES approach and ecosystem-based adaptation.

The ecosystem services approach

An ES approach is one that integrates the ecological, social and economic dimensions of natural resource management (Cork et al., 2007). Cork and colleagues (2007) have described an ES approach as the following.

- An ES approach helps to identify and classify the benefits that people derive from ecosystems. It also includes market and non-market, use and non-use, tangible and non-tangible benefits.
- It also explains consumers and producers of ES for maintenance and improvement of ecosystems for human well-being.
- This approach helps to describe and communicate benefits derived from natural and modified ecosystems to a wide range of stakeholders.

Ecosystem-based adaptation (EbA)

This approach integrates biodiversity and ES into an overall adaptation strategy to help people to adapt to the adverse effects of, for example, climate change (Colls et al., 2009). EbA can be applied at different geographical scales (local, regional, national) and over various periods (short to long term). It can be implemented as projects and as part of overall adaptation programmes. It is most effective when implemented as part of a broad portfolio of adaptation and development interventions (Colls et al., 2009). It is cost-effective and more accessible to rural or poor communities than measures based on hard infrastructure and engineering. It can integrate and maintain traditional and local knowledge and cultural values, such as in the New Zealand Maori concept of *Kaitiakitanga*.

This embraces the philosophy and practice of valuing inherited places and practices and aims to pass them on undamaged or improved. Some examples of EbA activities (CBD, 2009; Colls et al., 2009) are:

- coastal defence through the maintenance and/or restoration of mangroves and other coastal wetlands to reduce coastal flooding and coastal erosion;
- sustainable management of upland wetlands and floodplains for maintenance of water flow and quality;
- conservation and restoration of forests to stabilize land slopes and regulate water flows;
- establishment of diverse agroforestry systems to cope with increased risk from changed climatic conditions;
- conservation of agrobiodiversity to provide specific gene pools for crop and livestock adaptation to climate change.

Engineered systems

Engineered systems are landscapes such as farmland and cities that are actively modified to supply a particular set of ES. Farmland has been modified or 'engineered' to provide food and fibre, whereas cities have been actively managed to accommodate a human population. 'Engineered' or modified ecosystems are providers and consumers of different types of ES. Optimally managed 'engineered' or 'designed' ecosystems can provide a range of important ES; for instance, more fresh water, cleaner air and greater food production, as well as fewer floods and pollutants (Palmer et al., 2004). However, pursuit of commercial gains often reduces the ability to supply other vital ES. In this section and indeed in the following chapters, we discuss two modified or designed systems – agricultural and urban.

Agricultural systems

'Engineered' or modified ecosystems such as farmland are providers and consumers of different types of ES. Farmland comprises highly modified landscapes designed to generate revenue for farmers. Farmers use many inputs as well as natural inputs to produce food and fibre. The production of these is an ES. Intensive agriculture replaces many other ES with chemical inputs, resulting in a decrease in these services and their importance on farmland (Sandhu et al., 2008, 2010a, 2010b, 2012). This 'substitution agriculture' has to a large extent replaced these ES world-wide in the twentieth century. Severe environmental destruction, increasing fuel prices and the external costs of modern agriculture have resulted in increased interest among researchers and farmers in using ES for the more sustainable production of food and fibre (Daily, 1997; Costanza et al., 1997; Tilman, 1999; Cullen et al., 2004; Gurr et al., 2004, 2012; Robertson and Swinton, 2005). The above global trends have led to world-wide concerns about the environmental consequences of modern agriculture (Millennium Ecosystem

Assessment, 2005; De Schutter, 2010). There is also an additional concern that as the world approaches 'peak oil' and is already experiencing high oil prices, agriculture may no longer be able to depend so heavily on oil-derived 'substitution' inputs (Pimentel and Giampietro, 1994). Such a grave situation does not detract from the responsibility of agriculture to meet the food demands of a growing population but it does question its ability to increase yields without further ecosystem damage (Escudero, 1998; Tilman, 1999; Pimentel and Wilson, 2004; Schröter et al., 2005; UN, 2012). Therefore, the current challenge is to meet the food demands of a growing population and yet maintain and enhance the productivity of agricultural systems (UN, 1992). There is, therefore, currently an increasing interest in the services provided by nature.

It is now urgent that ES on farmland be enhanced as part of global food policy because increasingly dysfunctional biomes and ecosystems are appearing and agriculture, which largely created the problem, has become more intensive in its use of non-renewable resources, driven by a world population which is likely to reach nine billion people by 2050 (Foley et al., 2005). This intensification is compounded by a grain demand which is rising super-proportionally to human population increase and which is largely caused by biofuels development and a rapid rise in per capita meat consumption in parts of Asia (Rosegrant et al., 2001). Continuing with the current energy-intense (Pimentel et al., 2005), wasteful (Vitousek et al., 2009), polluting and unsustainable 'substitution agriculture', with its associated problems, which are likely to be exacerbated by climate change, is not an option for future world food security and productivity. There is, therefore, an urgent need for enhanced biodiversity-driven ES in world farming. Different types of agricultural systems and ES interactions are discussed in following chapters. More information is provided by Orre-Gordon et al., Sandhu et al. and Jordan and Warner in Chapters 4, 8 and 9, respectively. The relationship between aquaculture and ES is discussed in detail by Baulcomb in Chapter 5.

ES associated with agriculture

Costanza et al. (1997) estimated, with limited available data, the ES of world croplands to be only US$92 ha^{-1} year^{-1}. This was in marked contrast with other world biomes, for which ES were estimated to be worth US$23 000 ha^{-1} year^{-1} for estuaries, US$20 000 ha^{-1} year^{-1} for swamps and US$2000 ha^{-1} year^{-1} for tropical forests (Costanza et al., 1997). There are, however, two recent experimental agroecological approaches that can be used to demonstrate how this croplands figure can be much higher. The first involves agroecological experiments to measure ecosystem functions combined with value-transfer techniques to calculate their economic value. These studies demonstrate that some current farming practices have much higher ES values than in the Costanza et al. (1997) work. For example, recent data show that the combined value of only two ES (nitrogen mineralization and biological control of a single pest by one guild of invertebrate predators) can have values of US$197, $271 and $301 ha^{-1} year^{-1} in terms of avoided costs for conventional (Sandhu et al., 2008), organic (Lampkin, 1991) and integrated (Porter et al., 2009) arable farming systems, respectively. The above values comprise reduced variable costs (labour, fuel and pesticides) and lower external costs to human health and the environment. Paying for these variable costs is a charge to society, not to the individual farmer and although they contribute to GDP, that is a poor indicator of sustainability and of human well-being (Costanza, 2008).

The second recent realization that can transform ES on farmland is that a better understanding of ecological processes in agroecosystems can generate protocols which do not require a major farming system change but which enhance ES by returning selective functional agricultural biodiversity (FAB) to agriculture (Landis et al., 2000). For example, the role of leguminous crops in nitrogen fixation is a well-known enhancement of farmland ES and can have a value of US$40 ha^{-1} year^{-1} in terms of reduced oil-based fertilizer inputs (Vitousek et al., 2009), without including the value of reduced ES damage. More recent farmland ES improvements are illustrated by agroecological research on biological control of insect pests. In New Zealand and Australia, strips of flowering buckwheat *Fagopyrum esculentum* (Moench.) between vine rows provide nectar in an otherwise virtual monoculture and thereby improve the ecological fitness of parasitoid wasps that attack grape-feeding caterpillars. This in turn leads to the pest population being brought below the economic threshold. An investment of US$3 ha^{-1} year^{-1} in buckwheat seed and minimal sowing costs can lead to savings in variable costs of US$200 ha^{-1} year^{-1} as well as fewer pesticide residues in the wine, higher well-being for vineyard workers and enhanced ecotourism (Fountain and Tomkins, 2011).

Although the ecotechnologies now exist to improve farming sustainability when the negative consequences of oil-based inputs are well recognized, farmers world-wide are still largely risk averse (Anderson, 2003). They have traditionally rejected the idea that non-crop biodiversity on their land can improve production and/or minimize costs. The challenge now for agroecologists and policymakers is to use a range of market-based instruments or incentives, government interventions and enhanced social learning among growers to accelerate the deployment of sound, biodiversity-based ES-enhancement protocols for farmers. These protocols need to be framed in the form of service-providing units (Luck et al., 2003), which precisely explain the necessary ES-enhancement procedures and which should ideally include cost–benefit analyses. Such a requirement invites the design of new systems of primary production that ensure positive net carbon sequestration, are species diverse, have low inputs and provide a diverse suite of ES. An experimental example of such a system is a combined food, energy and ecosystem services (CFEES) agroecosystem in Denmark that uses non-food hedgerows as sources of biodiversity and biofuel. This novel production system is a net energy producer, providing more energy in the form of renewable biomass than is consumed in the planting, growing and harvesting of the food and fodder (Porter et al., 2009).

An approach to encouraging the uptake of ES-enhancing farming systems such as CFEES is through 'payment for ecosystem services' (PES) to private landowners (Food and Agriculture Organization, 2007). In this approach, those that benefit from the provision of ES make payments to those that supply them, thereby maintaining ES. Examples of working PES schemes currently in practice are found in different areas of the world. The current focus of these schemes is on water, carbon and biodiversity in addressing environmental problems through positive incentives to land managers (Food and Agriculture Organization, 2007). Such schemes would not only help to improve the environment and human well-being but also ensure food security and long-term farm sustainability (Rosegrant and Cline, 2003).

Although agricultural ecosystems may have low ES values per unit area when compared with others such as estuaries and wetlands, they offer the best chance of increasing global ES by developing appropriate goals for agriculture and the use of land management regimes that favour ES provision. This is because agriculture occupies 40% of the earth's land area and is readily amenable to changing practices, if the sociopolitical impediments are met. Agriculture can be considered to be the largest ecological experiment on Earth, with a high potential to damage global ES but also to promote them via ecologically informed approaches to the design of agroecosystems that value both marketed and non-marketed ES. The extensive Millennium Ecosystem Assessment (Millennium Ecosystem Assessment, 2005) of global ecosystems completed by science and policy communities provided a new framework for analysing socioecological processes and suggested that agriculture may be the 'largest threat to biodiversity and ecosystem function of any single human activity'. As 45% of the global population is engaged in farming activities, and such a large proportion of the global land area is in agriculture, achievement of human well-being as agreed by the UN-led Millennium Development Goals (MDGs) (UN, 2000) is not possible without clear pathways for the design of future agroecosystems. There are major global advantages of enhancing ES on farmland through adoption of ES-enhancement protocols. Therefore, global agricultural systems that utilize and maintain high levels of ES are required so that they can provide sustainable economic well-being and food security within ecological constraints (Royal Society, 2009). To condense this discussion into a simple goal, the farmer of the future needs to be encouraged to re-define his/her role to 'I am a photosynthesis manager and an ecosystem-service provider'.

Urban systems

Urbanization and urban growth are major drivers of ecosystem change globally. Urban areas are providing habitats for more than half the human population. In spite of these trends, the ecosystem idea has generally been applied to locations distant from the places where people live. However, knowledge about ecosystems is important for maintaining the quality of life in cities, suburbs and the fringes of metropolitan areas. Urban ecosystem concepts remind citizens and decision makers that we all ultimately depend on our ecosystems and their services (Daily, 1997). As the 'ecological footprint' of cities will increase in the coming decades, because they 'sequester' the products of ES from elsewhere, there is need to incorporate ES into decision making during planning and management of urban areas.

Urban ecosystems have been neglected due to the lack of understanding of the complex processes involved, the lack of mechanisms to govern them, and the failure to incorporate ES into day-to-day decision making. Urban development trends pose serious problems with respect to ES and human well-being. The Millennium Ecosystem Assessment (2005) treated urban systems as ecosystems necessary for human welfare. As they are dominated by humans, these systems can be classified on the basis of population size, economic condition and location. Nearly half the world's population lives in cities of less than half a million people and about 10% lives in those with more than 10 million (Millennium Ecosystem Assessment, 2005). The ES challenges within cities are enormous and are discussed in this chapter below and later in this book.

ES in urban systems

Urban systems are not functional or self-contained ecosystems. They depend largely on surrounding ecosystems in rural areas or more distant ecosystems to fulfil their daily needs including food, water and material for housing and other needs. In cities, urban parks, forests and green belts have their strategic importance for the quality of life. They provide essential ES such as gas regulation, air and water purification, wind and noise reduction, etc. They also enhance social and cultural services such as feelings of well-being, and provide recreational opportunities for urban dwellers (Miller, 1997; Smardon, 1988; Botkin and Beveridge, 1997; Bolund and Hunhammar, 1999; Lorenzo et al., 2000; Tyrväinen and Miettinen, 2000).

Towns and cities are also both consumers and producers of ES. However, the net flow of ES is invariably into rather than out of urban systems. Even if they are not major producers of ES, urban activities can alter the supply and flow of ES at every scale, from local to global level. Urban development threatens the quality of the air, the quality and availability of water, the waste processing and recycling systems, and many other qualities of the ambient environment that contribute to human well-being.

ES and their interactions in engineered systems

Both agricultural and urban systems are dependent and impact on the provision of ES. These designed systems are affected by direct and indirect drivers that in turn impact ES (Fig. 1.1). It is very important to understand these interactions between ES and 'engineered systems' for the achievement of equitable and sustainable human welfare (Swaminathan, 2012).

Human society, as part of the planetary system of interacting biomes depends on these ES as life support functions. Yet simultaneously we are impacting negatively on ecosystem goods and services. This is the dilemma facing society as our ecological footprint on planet earth increases. Projected economic expansion to meet the demands of a growing population (projected to be 9 billion by 2050) along with global climate change will jeopardize future human well-being by further degrading ecosystems. There is a great need to incorporate the value of ES into day-to-day decision making, into government policies and in business practices so that sustainable and desirable futures can be achieved. Waste of energy, food and other resources in the 'developed' world points to areas where our current practices can be readily modified.

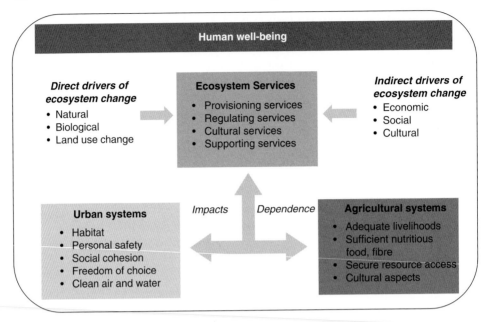

Fig. 1.1   Framework of drivers of ecosystem change and the interaction between ES and two 'engineered systems' – urban and agricultural systems.

In this context, global studies have largely focused on natural ecosystems and biomes, such as the boreal forests and the sea and have put little emphasis on managed ecosystems such as farmland and cities. However, the continued supply of ecosystem goods and services is of vital significance for the survival and productivity of our farmland and our cities. Agricultural systems comprise the largest managed ecosystems on Earth, and are often confronted by ecosystem degradation. Much of the success of modern agriculture has been from provisioning services such as food and fibre. However, the expansion in the demand and supply of these marketable ecosystem goods has resulted in the suppression of other valuable and essential ES such as pollination, climate and water regulation, biodiversity and soil conservation. Similarly, demands from urban areas to support and enhance human lifestyles have resulted in the degradation of other valuable ES in other parts of the world. As economic wealth is underpinned by ecological wealth, we need to recognize and understand the role of ES in sustaining societies, nations and individuals. This can help to achieve food security and environmental sustainability at scales from local to global. It can help ensure a sustainable development and an equitable future. Without the evaluation, protection and enhancement of ES in agriculture and cities, the world's future is bleak indeed.

References

Anderson, J.R. (2003). Risk in rural development: challenges for managers and policy makers. *Agricultural Systems*, 75, 161–197.

Balmford, A., Fisher, B., Green, R.E., et al. (2011). Bringing ecosystem services into the real world: an operational framework for assessing the economic consequences of losing wild nature. *Environmental and Resource Economics*, **48**, 161–175.

Bolund, P. and Hunhammar, S. (1999). Ecosystem services in urban areas. *Ecological Economics*, **29**, 293–301.

Botkin, D.B. and Beveridge, C.E. (1997). Cities as environments. *Urban Ecosystems*, **1**, 3–19.

Boyd, J. and Banzhaf, S. (2007). What are ecosystem services? *Ecological Economics*, **63**, 616–626.

CBD (2009). *Connecting Biodiversity and Climate Change Mitigation and Adaptation*. Report of the Second Ad Hoc Technical Expert Group on Biodiversity and Climate Change. Montreal, Technical Series No. 41.

Colls, A., Ash, N. and Ikkalal, N. (2009). *Ecosystem Based Adaptation: a Natural Response to Climate Change*. IUCN Report, p. 16.

Cork, S., Stoneham, G. and Lowe, K. (2007). *Ecosystem Services and Australian Natural Resource Management (NRM) Futures*. Paper to the Natural Resource Policies and Programs Committee (NRPPC) and the Natural Resource Management Standing Committee (NRMSC). Australian Government Department of the Environment, Water, Heritage and the Arts, Canberra, Australia.

Costanza, R. (2008). Stewardship for a 'full' world. *Current History*, **107**, 30–35.

Costanza, R., d'Arge, R., De Groot, R., et al. (1997). The value of the world's ecosystem services and natural capital. *Nature*, **387**, 253–260.

Cullen, R., Takatsuka, Y., Wilson, M. and Wratten, S. (2004). *Ecosystem Services on New Zealand Arable Farms*. Agribusiness and Economics Research Unit, Lincoln University, Discussion Paper **151**, pp. 84–91.

Daily, G.C. (1997). *Nature's Services: Societal Dependence on Natural Ecosystems*. Island Press, Washington, DC.

Daily, G.C., Alexander, S., Ehrlich, P.R., et al. (1997). Ecosystem services: benefits supplied to human societies by natural ecosystems. *Issues in Ecology*, **2**, 18.

de Groot, R.S. (1992). *Functions of Nature: Evaluation of Nature in Environment Planning, Management and Decision Making*. Wolters-Noordhoff, Gröningen.

de Groot, R.S., Wilson, M. and Boumans, R.M.J. (2002). A typology for the classification, description and valuation of ecosystem functions, goods and services. *Ecological Economics*, **41**, 393–408.

De Schutter, O. (2010). *Agroecology and the Right to Food, Report submitted by the Special Rapporteur on the Right to Food*, p.21. United Nations General Assembly.

Escudero, G. (1998). The vision and mission of agriculture in the year 2020: towards a focus that values agriculture and the rural environment. Agricultura, Medioambiente y Pobreza Rural en America Latina (eds L.G. Reca and R.G., Echeverria), pp. 21–54. CAB Direct, Washington, DC.

Fisher, B. and Turner, K. (2008). Ecosystem services: classification for valuation. *Biological Conservation*, **141**, 1167–1169.

Fisher, B., Turner, K., Zylstra, M., et al. (2008). Ecosystem services and economic theory: integration for policy-relevant research. *Ecological Applications*, **18**, 2050–2067.

Foley, J.A., deFries, R., Asner, G.P., et al. (2005). Global consequences of land use. *Science*, **309**, 570–573.

Food and Agriculture Organization (2007). *The State of Food and Agriculture: Paying Farmers for Environmental Services*, p. 222.

Fountain, J. and Tompkins, J. (2011). The potential of wine tourism experiences to impart knowledge of sustainable practices: the case of the Greening Waipara biodiversity trails. *Proceedings of the 6th AWBR International Conference*, 9–10 June, Bordeaux Management School, France.

Gurr, G.M., Wratten, S.D. and Altieri, M.A. (eds) (2004). *Ecological Engineering for Pest Management: Advances in Habitat Manipulation for Arthropods*. CSIRO, Victoria.

Gurr, G.M., Wratten, S.D., Snyder, W.E. and Read, D.M. (2012). *Biodiversity and Insect Pests – Key Issues for Sustainable Management*. Wiley-Blackwell, UK.

IPBES (2010). *Intergovernmental Science-Policy Platform on Biodiversity and Ecosystem Services.* UNEP. Available at http://ipbes.net/ (accessed August 2012).

Lampkin, N. (1991). *Organic Farming.* Farming Press, Ipswich, UK.

Landis, D.A., Wratten, S.D. and Gurr, G.M. (2000). Habitat management to conserve natural enemies of arthropod pests in Agriculture. *Annual Review of Entomology,* **45**, 175–201.

Lorenzo, A.B., Blanche, C.A., Qi, Y. and Guidry, M.M. (2000). Assessing residents' willingness to pay to preserve the community urban forest: a small-city case study. *Journal of Arboriculture,* **26**, 319–325.

Losey, J.E. and Vaughan, M. (2006).The economic value of ecological services provided by insects. *BioScience,* **56**, 311–323.

Luck, G.W., Daily, G.C. and Ehrlich, P.R. (2003). Population diversity and ecosystem services. *Trends in Ecology and Evolution,* **18**, 331–336.

Millennium Ecosystem Assessment (2005). *Synthesis Report.* Island Press, Washington, DC.

Miller, R. W. (1997). *Urban Forestry – Planning and Managing Urban Greenspaces,* 2nd edn. Prentice-Hall, Englewood Cliffs, NJ.

Mooney, H., Larigauderie, A., Cesario, M., et al. (2009). Biodiversity, climate change and ecosystem services. *Current Opinion in Environment Sustainability,* **1**, 46–54.

Myers, N. (1996). Environmental services of biodiversity. *Proceedings of the National Academy of Sciences USA,* **93**, 2764–2769.

Palmer, M., Bernhardt, E., Chornesky, E., et al. (2004). Ecology for a crowded planet. *Science,* **304**, 1251–1252.

Pimentel, D., Hepperly P., Hanson, J., Douds, D. and Seidel, R. (2005). Environmental, energetic, and economic comparisons of organic and conventional farming systems. *Bioscience,* **55**, 573–582.

Pimentel, D. and Giampietro, M. (1994). *Food, Land, Population and the U.S. Economy.* Carrying Capacity Network, Washington, DC.

Pimentel, D. and Wilson, A. (2004). World population, agriculture and malnutrition. *World Watch,* **17**, 22–25.

Porter, J.R.P., Costanza, R., Sandhu, H., Sigsgaard, L. and Wratten, S. (2009). The value of producing food, energy, and ecosystem services within and agro-ecosystem. *Ambio,* **38**, 186–193.

Robertson, G.P. and Swinton, S.M. (2005). Reconciling agricultural productivity and environmental integrity: a grand challenge for agriculture. *Frontiers in Ecology and Environment,* **3**, 38–46.

Rosegrant, M.W. and Cline, S.A. (2003). Global food security: challenges and policies. *Science,* **302**, 1917–1919.

Rosegrant, M.W., Paisner, M., Meijer, S. and Whitcover, J. (2001). *Global Food Projections to 2020: Emerging Trends and Alternative Futures.* IFPRI, Washington, DC.

Royal Society (2009). *Reaping the Benefits: Science and the Sustainable Intensification of Global Agriculture.* Available at http://royalsociety.org/Reapingthebenefits/ (2009), p. 86 (accessed August 2012).

Sagoff, M. (2011). The quantification and valuation of ecosystem services. *Ecological Economics,* **70**, 497–502.

Sandhu, H., Crossman, N. and Smith, F. (2012). Ecosystem services in Australian agricultural enterprises. *Ecological Economics,* **74**, 19–26.

Sandhu, H.S., Wratten, S.D. and Cullen, R. (2010a). Organic agriculture and ecosystem services. *Environmental Science and Policy,* **13**, 1–7.

Sandhu, H.S., Wratten, S.D. and Cullen, R. (2010b). The role of supporting ecosystem services in conventional and organic arable farmland. *Ecological Complexity,* **7**, 302–310.

Sandhu, H.S., Wratten, S.D., Cullen, R. and Case, B. (2008).The future of farming: the value of ecosystem services in conventional and organic arable land. An experimental approach. *Ecological Economics,* **64**, 835–848.

Schröter, D., Cramer, W., Leemans, R., et al. (2005). Ecosystem service supply and vulnerability to global change in Europe. *Science,* **310**, 1333–1337.

Smardon, R.C. (1988). Perception and aesthetics of the urban environment: review of the role of vegetation. *Landscape and Urban Planning*, **15**, 85–106.

Swaminathan, M.S. (2012). *Sustainable Development: Twenty Years after Rio*. International Consultation on Twenty years of Rio: Biodiversity-Development-Livelihoods, Chennai, India.

Tilman, D. (1999). Global environmental impacts of agricultural expansion: The need for sustainable and efficient practices. *Proceedings of the National Academy of Sciences USA*, **96**, 5995–6000.

Toman, M. (1998). Why not to calculate the value of the world's ecosystem services and natural capital. *Ecological Economics*, **25**, 1, 57–60.

Tyrväinen, L. and Miettinen, A. (2000). Property prices and urban forest amenities. *Journal of Environmental Economics and Management*, **39**, 205–223.

UN (1992). *Promoting Sustainable Agriculture and Rural Development. United Nations Conference on Environment and Development*. Rio de Janeiro, Brazil, 3 to 14 June. Agenda 21, 14.1-14.104. Available at http://www.un.org/esa/sustdev/agenda21.htm (accessed August 2012).

UN (2000). *United Nations Millennium Declaration*. Available at: http://www.un.org/millennium/declaration/ares552e.pdf (accessed August 2012).

UN (2012). *The Future We Want. United Nations Conference on Sustainable Development*. United Nations, Rio de Janeiro.

Vitousek, P.M., Naylor, R., Crews, T., et al. (2009). Nutrient imbalances in agricultural development. *Science*, **324**, 1519–1520.

Wallace, K.J. (2007). Classification of ecosystem services: problems and solutions. *Biological Conservation*, **139**, 235–246.

2

Ecological Processes, Functions and Ecosystem Services: Inextricable Linkages between Wetlands and Agricultural Systems

Onil Banerjee,[1] Neville D. Crossman[1] and Rudolf S. de Groot[2]

[1] CSIRO Ecosystem Sciences, Adelaide, Australia
[2] Environmental Systems Analysis Group, Wageningen University, The Netherlands

Abstract

Ecosystems contribute to human well-being via the provision of goods and services where the benefits are direct, such as in the production of food and raw materials, and indirect as is the case in the regulation of water quality and supply. Underpinning these services is a suite of ecological functions that must be understood in order to manage and enhance ecosystem services provision. For example, a healthy wetland that contains a biologically diverse array of producers and consumers purifies water, making freshwater available for irrigated agricultural production, which in turn provides food for human consumption. Making the link between function and service also enables us to identify threats to ecosystem services from unsustainable management practices. For example, the excessive use of chemicals in agricultural production affects water quality and threatens a wetland's functional capacity to purify water, consequently affecting food production. In this chapter, we identify the relationships between ecosystem function and ecosystem service. This linkage is a precursor to the estimation of ecosystem service values and understanding how changes in land and water management flow through to marginal changes in values. To contextualize this relationship, we consider specifically the services that wetlands provide in support of agricultural systems. We conclude with research challenges on managing complexity, resilience and trade-offs between ecosystem services and agriculture.

Ecosystem Services in Agricultural and Urban Landscapes, First Edition. Edited by Steve Wratten, Harpinder Sandhu, Ross Cullen and Robert Costanza.
© 2013 John Wiley & Sons, Ltd. Published 2013 by John Wiley & Sons, Ltd.

Introduction

A critical challenge in the integration of ecosystem and economic science is the development of an operational classification of ecosystems and their functions which lends itself to the valuation of ecosystem services (de Groot et al., 2002; National Research Council, 2005). In the absence of either a political mandate to protect ecosystem integrity or a method of assigning value to ecosystem services for use in decision making, land use and development decisions will continue to be made without sufficient consideration for the important role ecosystems play in sustaining life (National Research Council, 2005; Daily et al., 2009). Furthermore, assigning monetary value to ecosystem services can aid in making environmental problems visible and thus inform decision processes (Wilson and Howarth, 2002; Spangenberg and Settele, 2010).

The provision of ecosystem services and subsequent benefit to humans is underpinned by a series of biophysical processes and ecological functions which themselves are driven by biological diversity (Balvanera et al., 2006). These linkages are highlighted in Fig. 2.1. Experiments have shown that increasing the amount of biological diversity has in most cases an increasingly positive effect on ecosystem function and service. For example, greater abundance of soil mycorrhiza and a higher rate of soil decomposer activity increases the rate of nutrient cycling, which is a regulating ecosystem service. A faster rate of nutrient cycling can be of direct benefit to humans if harnessed to increase agricultural productivity.

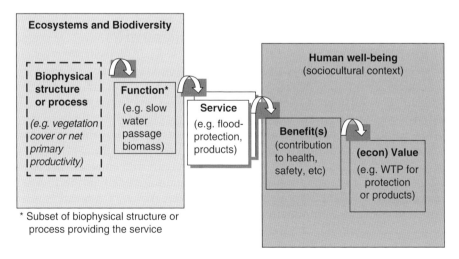

Fig. 2.1 The interdependencies of biological diversity, biophysical process, ecosystem function and service, human well-being, and willingness to pay (WTP). From de Groot, R.S., Alkemade, R., Braat, L., Hein, L. and Willemen, L. (2010). Challenges in integrating the concept of ecosystem services and values in landscape planning, management and decision making. *Ecological Complexity*, **7**, 260–272.

Agricultural commodities, valued in the market place, are just one of the ecosystem services agricultural systems produce. Ecosystem services have use and non-use values, and are valued using various methods. Non-use values include existence, bequest and altruistic values, or simply put – the knowledge that an ecosystem exists for us and for others now and in the future is valuable (National Research Council, 2005; Turner et al., 2008). Use values are categorized as direct and indirect. Direct-use values include timber production; a scenic lake may have recreational value which is captured by a management authority, or a home with a view of a natural and structurally diverse forest may fetch a better market price than a similar house without a scenic view. Ecosystems generate a multitude of indirect use values such as water filtration, nutrient retention and erosion mitigation. These values are less tangible than direct-use values and do not directly involve interaction between a beneficiary and the ecosystem (TEEB, 2010).

In this chapter we document the relationship between biological diversity, ecosystem function and service within agricultural systems. To guide the discussion, we focus on the interdependencies between agricultural production and the ecosystem services provided by freshwater wetlands (hereafter *wetlands*) and the impacts agricultural systems can have on the health and functioning of wetlands. We focus on wetlands because they are biologically complex yet relatively well understood, and critical to the provision of freshwater for agricultural use and human benefit. In the section that follows, ecosystem function and its linkages with ecosystem services are established. The ecological functions and subsequent ecosystem services generated by wetlands are defined and their interactions with agricultural systems are discussed in detail. We conclude the chapter with a discussion of the research challenges involved in managing complexity, resilience and trade-offs between ecosystem services and agriculture.

Linking ecosystem function with ecosystem service

Ecosystems directly contribute to human well-being via the provision of ecosystem services (Costanza et al., 1997; Daily, 1997; Millennium Ecosystem Assessment, 2003; Perrings, 2006; TEEB, 2010). The benefits provided by ecosystem services within agricultural systems are direct, such as food and raw materials, and indirect and include the regulation of water supply and quality and nutrient cycling example. Underpinning these services is a suite of ecological functions that must be understood in a first step to valuing, managing and enhancing ecosystem service provision. Importantly, a healthy and functioning wetland purifies water via biogeochemical and nutrient-retention processes, making freshwater available for irrigated agricultural production, which in turn provides food for human consumption. Making the link between function and service also enables us to identify threats to ecosystem services from unsustainable management practices. For example, agricultural run-off that follows from excessive pesticide or fertilizer use impedes biogeochemical and nutrient retention processes, threatening the ability of wetlands to purify water, which in turn threatens food production.

Ecosystem functions result from the interactions between characteristics, structures and processes (Turner et al., 2000) constituting the physical, chemical and biological exchanges and processes that contribute to the self-maintenance and self-renewal of an ecosystem (e.g. nutrient cycling and food-web interactions). Ecosystem functions involve interactions between biotic and abiotic system components in achieving any and all ecosystem outcomes (National Research Council, 2005). de Groot (1992) illustrates the link between ecosystem function and human benefit by defining function as the capacity of natural processes and components to provide goods and services that generate human utility. Linking ecosystem function to human benefit should encourage ecosystem-based management because of the monetary or non-monetary benefits provided by functionally diverse systems (Turner et al., 2008; Willemen et al., 2010).

Following the Millennium Ecosystem Assessment (Millennium Ecosystems Assessment, 2005), ecosystem functions may be conveniently grouped into four categories, namely: production, regulation, habitat and informational functions. Regulatory functions include gas and nutrient exchange, disturbance prevention, water regulation, soil retention and formation, waste treatment, pollination and biological control. Critical habitat functions are the provision of habitat and maintenance of biological diversity, while the production function includes the production of food and other raw materials such as medicinal, genetic and ornamental resources. Informational functions include aesthetic, recreational, cultural and spiritual functions.

Ecosystem function and their resulting services have an inherently spatial nature. Services may be created and the benefits enjoyed in situ. An example of this is the provision of habitat which may be used by animals that are subsequently hunted for recreation. Benefits may be omnidirectional where services are created in one location, though the benefits are spatially extensive, which is the case of the role of wetlands in sequestering carbon (Zedler and Kercher, 2005) and thus mitigating climate change – a benefit enjoyed globally. Finally, services may be directional, where a function occurs in one location, while the benefits are perceived directionally from that location due to the direction of flow. An example of this is the function riparian ecosystems serve in downstream flood control (Zedler and Kercher, 2005; Turner et al., 2008).

Wetlands

Wetlands are particularly diverse and productive ecosystems (Woodward and Wui, 2001; Zedler and Kercher, 2005) providing direct and indirect benefits at local, landscape and global scales (Acharya, 2000). Wetlands may be defined as areas exhibiting a temporary or permanent presence of water above or close to the soil surface and are maintained by waterlogging. Water is the primary factor affecting plant and animal life in these systems. Wetlands, although occupying less than 9% of the earth's terrestrial surface, contribute significantly in the provision of ecosystem services (Zedler and Kercher, 2005).

There are three major types of freshwater wetlands (Barbier et al., 1997): riverine, palustrine and lacustrine wetlands. Riverine wetlands are areas that are periodically flooded by a river rising above its banks and include water meadows, flooded forests and oxbow lakes. Palustrine wetlands are characterized by a mostly permanent presence of water and include ponds and kettle and volcanic crater lakes. Lacustrine wetlands are permanently inundated areas with minimal water flow. The following sections provide an overview of key wetland functions, linkages to ecosystem services and their relationship with agricultural systems.

Wetland functions

Wetlands provide regulation (hydrological and biogeochemical), production, habitat and informational functions. The hydrological aspects of a wetland are critical in defining their characteristics and processes (Maltby, 2009). Three principal hydrological functions of wetlands are floodwater detention, groundwater recharge/discharge and sediment retention (Turner et al., 2008). Table 2.1 describes the linkages between wetland function and ecosystem service, and presents metrics to assess the presence and level of service provision.

A wetland's hydrological function contributes to its high productivity through the capture and cycling of nutrients from upstream (Barbier et al., 1997). Wetlands reduce overbank flooding and slope run-off (Zedler and Kercher, 2005). By storing water, wetlands delay and reduce peak flows which could otherwise cause downstream flood damage. Wetlands may have significant interactions with groundwater where the substrate between the two is permeable. In these cases, wetlands may be involved in groundwater recharge and/or discharge of aquifers (Maltby, 2009). Finally, wetlands serve to retain sediments thereby alleviating downstream navigational problems, water treatment costs and damage to pumping infrastructure and spawning habitat.

The interaction of a wetland's biogeochemical function with hydrological functions enables interactions with surrounding wetlands (Mander et al., 2005). Specifically, biogeochemical functions of wetlands influence water quality, pollution control and biodiversity (Mander et al., 2005; Zedler and Kercher, 2005; Maltby, 2009). Oxidization and reduction processes in the soil are responsible for significant biogeochemical reactions. Wetland flooding results in oxygen depletion where, through time, organic substrates are consumed and oxygen, nitrates and other compounds are reduced. The inundation of floodplains facilitates nutrient exchange; these sites are also often important spawning grounds for fish.

The nutrient retention function of wetlands can affect water quality considerably, especially through the mitigation of incoming pollution. Nutrients and trace elements may be retained in plant structures or soil and organic matter (Mander et al., 2005), while nutrient export contributes to water quality maintenance and occurs through gaseous emission (Zedler, 2003), biomass harvest or erosion. Carbon is also retained in wetlands and is dependent on waterlogging, pH, nutrients and temperature. The level of pH and aerobic conditions in a wetland affects biodiversity in terms of the species and community assemblages possible. Organic carbon concentrations affect water turbidity and pH (Maltby, 2009).

With regards to habitat function, wetlands often support a disproportionately large amount of biodiversity, including a significant number of rare or endangered species. Efforts aimed at protecting wetlands are often driven by concern for their biodiversity (Zedler and Kercher, 2005). A higher level of species diversity is promoted by ecological disturbance that occurs as a consequence of wetting

Table 2.1 Wetland ecosystem function, service and indicator.

Ecosystem function	Ecosystem service	Establishing presence	State indicator; sustainable yield
Provisioning			
	Food	Fish, game, fruits and grains	Total or average stock (kg ha^{-1}) Net productivity (Kcal year^{-1})
	Water	Water storage for domestic/ industrial/ agricultural use	Total water (cubic m ha^{-1}) Net water inflow (m^3 year^{-1})
	Fibre, fuel and other raw material	Biotic/ abiotic resources, e.g. peat, fodder, fuel wood	Total biomass (kg ha^{-1}) Net productivity (kg year^{-1})
	Genetic resources	Genes for pathogen resistance, ornamental species	Number of species Maximum sustainable harvest (kg ha^{-1})
	Biochemical and medicinal resources	Potential medicines and other biotic materials	Amount of useful substances (kg ha^{-1}) Maximum sustainable harvest (kg ha^{-1})
Regulating			
	Air quality	Capacity to extract atmospheric aerosols and chemicals	Leaf Area Index or NOx-fixation Quantity of aerosols/ chemicals extracted
	Climate	Influence on global and local climate	Greenhouse gas balance, carbon sequestration, land cover Quantity of GHGs fixed
	Water regulation	Groundwater recharge/ discharge	Surface or soil water retention capacity Quantity of water stored and influence of hydrological regime
	Waste treatment	Biotic and abiotic processes to remove excess nutrients/ pollutants	Denitrification (kg N ha^{-1} year^{-1}) Immobilization in plants and soil Maximum amount of waste recycled and influence on water and soil parameters
	Erosion protection	Soil and sediment retention	Root matrix Amount of soil/ sediment captured/ retained
	Soil formation and regeneration	Natural processes in soil formation and regeneration	Bioturbation
	Pollination	Habitat for pollinators	Number and impact of pollinating species

(continued)

Table 2.1   *(Cont'd)*

Ecosystem function	Ecosystem service	Establishing presence	State indicator; sustainable yield
	Biological regulation	Control of pests through trophic relations	Number and impact of pest-control species Reduction of disease and pests, and crop pollination dependence
	Natural hazard	Forests and dampening extreme events	Water storage in cubic meters Reduction of flood danger and prevention of infrastructure damage
Habitat			
	Nursery	Breeding, feeding and resting habitat	Number of species and individuals Ecological value
	Gene pool	Maintenance of ecological balance	Natural biodiversity; endemic species Habitat integrity
Information			
	Aesthetic	Structural diversity and other factors	Number/area of landscape features Number of sustainable users
	Recreational and inspirational	Landscape features	Number/area of landscape features Number of sustainable users
	Cultural	Culturally significant features	Number/area or presence of landscape features Number of users
	Spiritual	Spiritually significant features	Number/area or presence of landscape features Number of users

Sources: de Groot et al. (2002); de Groot et al. (2006); Food and Agriculture Organization (2008).

and drying cycles of wetlands. The production function of wetlands involves the conversion of energy, nutrients, water and gases into living biomass. This is a form of food-web support – the efficient primary production of biomass (Maltby, 2009). This function generates significant human utility through its production and provision of raw materials. Wetlands also serve an important function in maintaining habitat connectivity (Zedler, 2003; Mander et al., 2005; Tscharntke et al., 2005). Finally, information functions contribute to human cognitive, emotional and spiritual health, among other things.

Wetland–agricultural systems interactions

Agricultural systems rely on ecosystem services to enable the production of food, fibre, bioenergy and pharmaceuticals, and other important commodities. This present volume as well as recent research discuss in detail the ecosystem services on which agriculture depends (Porter et al., 2009; Power, 2010; Ribaudo et al., 2010; Sandhu et al., 2010a, 2010b, 2012). Approximately 20% of global agriculture depends on blue water (i.e. freshwater) extracted from surface water and groundwater resources and close to 70% of global water withdrawal is used for agricultural purposes (Comprehensive Assessment of Water Management in Agriculture, 2007). The water filtration service undertaken by wetlands is therefore critical to agricultural productivity.

In addition to ensuring adequate water quality and supply, wetlands provide agriculture with services related to pollination, biological pest control, maintenance of soil structure and fertility, and erosion mitigation. Wetlands mitigate floods and reduce floodwater peaks; they replenish stream flow through subsurface flow, contribute to water table recharge and, depending on their position in the landscape, wetlands may retain water from aquifer discharge (Food and Agriculture Organization, 2008). Wetlands and riparian areas influence microclimates of adjacent fields by regulating humidity and evapotranspiration, and serve in filtering often contaminated overland flow from intensively managed agricultural areas (Mander et al., 2005).

Various crops such as rice, corn, some vegetables and fruits are grown in, or in proximity to, wetlands. Activities such as fishing, livestock grazing and hay production are also conducted in or supported by these ecosystems. Soils in these areas are typically quite fertile with high clay content, particularly in seasonally inundated floodplains (Food and Agriculture Organization, 2008). Agricultural systems themselves produce ecosystem services (Tscharntke et al., 2005): they sequester carbon, regulate soil fertility, retain and cycle nutrients, and provide landscapes with aesthetic, cultural and spiritual values (Antle and Stoorvogel, 2006; Porter et al., 2009; Ribaudo et al., 2010). Wetlands support not only agriculture in these ways, but also agricultural communities, by providing potable water and adequate supply for hydroelectric power generation. Wetlands and agricultural systems are therefore inextricably linked as they provide agriculture with critical and valuable services.

Negative feedbacks, otherwise known as disservices (Power, 2010), created by agricultural systems have adverse impacts on wetlands through habitat deterioration, contamination of fisheries and spawning areas, biodiversity loss, run-off, sedimentation, greenhouse gas emissions and the release of toxins into the environment. The primary pathway by which agricultural systems affect wetlands is through the diversion of water for irrigation and nutrient loading of nitrogen and phosphorous (Millennium Ecosystems Assessment, 2005; Comprehensive Assessment of Water Management in Agriculture, 2007).

Irrigated agriculture in some regions has resulted in soil salinization, equating to a global loss of 1.5 million hectares of arable land per year. Furthermore, large quantities of salt from land salinization are transported into wetlands by irrigation run-off, having substantial impacts on biodiversity, productivity and biogeochemical composition in wetlands (Williams, 2001). Changes to water regimes can have devastating effects on wetlands and their regulating functions including those dependent on groundwater, surface water and direct rainfall. Wetland degradation may expose agricultural systems to increased vulnerability to storm, flood and eutrophication events.

The interactions between wetlands and agricultural systems may be characterized as in situ or external where the former constitutes an agricultural intervention within a wetland and the latter is an intervention that is upstream, downstream or peripheral to a wetland. In situ interactions may involve a substantial transformation of the wetland ecosystem or a more benign interaction. Significantly altering the ecosystem could involve drainage, grazing, ploughing or the application of pesticides and fertilizers. Fishing or the managed gathering of plants and animals is considered non-transformative, while enhancement can include manipulation of wetlands for agricultural or aquacultural purposes, including the creation of rice paddies, fish ponds and water storage areas (Food and Agriculture Organization, 2008).

External interactions are more common than direct wetland interventions. Upstream interactions can involve diversion of water to agriculture which may have water quantity, quality and flow effects to wetlands situated downstream. Return flows of diverted water will be lower in quantity and may contain substantial amounts of nutrients and toxins. Hydraulic gradients may also be created resulting in more rapid release of upland water and a lower watertable. Upstream agricultural practices that create erosion, sedimentation and runoff are detrimental to wetland ecosystems (Zedler and Kercher, 2005). Less common is the case where wetlands affect agricultural activity upstream through their capacity for water storage and sediment retention; should their capacity in this regard be compromised, upstream waterlogging of agricultural areas may result (Food and Agriculture Organization, 2008). Furthermore, these types of interactions are seldom confined to one agricultural production unit and wetland, rather these interactions typically occur and are compounded at the catchment scale.

Some research challenges

Understanding complexity and resilience

Ecosystems provide numerous goods and services, many of which have indirect value and are not traded in the market place. Our understanding of the ecosystem functions underpinning these services is limited, complicated by the spatial and temporal scales over which ecosystem services operate, and the interdependencies between ecosystem components and functions. Ecosystem functions are dynamic, exhibiting thresholds, complementary relationships to keystone processes, and system integrity and irreversibility (Turner et al., 2008). A threshold occurs where an ecosystem may cease to function or may function in an alternative undesirable state because one or more of its attributes are degraded beyond a specific level. Complementary relationships describe the interactions and interdependence of ecosystem components where the survival of one species depends on the existence of other species. These relationships have contributory value, which is a reflection of limited substitution possibilities. The notion of keystone processes describes system dependence on a limited number of ecosystem functions. A reduction in ecosystem diversity (e.g. structural or species diversity) can affect system resilience and adaptability to shocks. Ecosystem structure and function reflects the notion that the health of an ecosystem depends on system integrity and the whole functioning of the system.

Trade-offs

Management and planning for wetlands and agriculture should focus on enhancing multifunctionality where multiple ecosystem services are provided for human well-being and economic development. There is great potential to achieve synergies and win–win outcomes from effective planning and the development of economic incentives (DeFries and Rosenzweig, 2010; Gordon et al., 2010; Raudsepp-Hearne et al., 2010). However, the less desirable lose–lose or lose–win outcomes are commonplace due to trade-offs between services and agriculture production (Tallis et al., 2008; Gordon et al., 2010; Crossman et al., 2011). Trade-offs arise when provisioning services, especially agricultural production, seem to conflict with regulating, habitat and information services. Globally, most wetland ecosystems have been heavily modified to make way for food provisioning at the expense of other ecosystem services (Comprehensive Assessment of Water Management in Agriculture, 2007). The principle cause for the decline of ecosystem services other than provisioning services, and a major barrier to the evolution of multifunctional landscapes, is the lack of economic valuation of these services. Where the value of these services is not accounted for in decision-making frameworks, such as cost–benefit analysis, the importance of these services in support of agricultural production are overlooked and trade-offs may be made using poor information.

Management of wetlands and surrounding agricultural landscapes needs to account for the values of multiple ecosystem services (Carpenter et al., 2009). While there are an increasing number of examples of the creation of markets for ecosystem goods and services, including the provision of freshwater (Carroll et al., 2008; Bayon et al., 2009; Garrick et al., 2009), markets for most services are either absent or immature, leading to a lack of appropriate price signals for enhancing multifunctionality. Major challenges that lie ahead are the design of efficient markets for ecosystem service provision, and the development of strong institutions and regulatory instruments that underpin these markets. The goal is the sustainable growth of agricultural provisioning services without increasing the production of ecosystem disservices as these markets and institutions evolve.

References

Acharya, G. (2000). Approaches to Valuing the hidden hydrological services of wetland ecosystems. *Ecological Economics*, 35, 63–74.

Antle, J.M. and Stoorvogel, J.J. (2006). Predicting the supply of ecosystem services from agriculture. *American Journal of Agricultural Economics*, 88, 1174–1180.

Balvanera, P., Pfisterer, A.B., Buchmann, N., et al. (2006). Quantifying the evidence for biodiversity effects on ecosystem functioning and services. *Ecology Letters*, 9, 1146–1156.

Barbier, E.B., Acreman, M. and Knowler, D. (1997). *Economic Valuation of Wetlands. A Guide for Policy Makers and Planners*. Ramsar Convention Bureau, Gland.

Bayon, R., Hawn, A. and Hamilton, K. (2009). *Voluntary Carbon Markets: An International Business Guide to What They Are and How They Work*. Earthscan, London.

Carpenter, S.R., Mooney, H.A., Agard, J., et al. (2009). Science for managing ecosystem services: beyond the Millennium Ecosystem Assessment. *Proceedings of the National Academy of Sciences*, **106**, 1305–1312.

Carroll, N., Fox, J. and Bayon, R. (eds) (2008). *Conservation and Biodiversity Banking: A Guide to Setting up and Running Biodiversity Credit Trading Systems*. Earthscan, London.

Comprehensive Assessment of Water Management in Agriculture (2007). *Water for Food, Water for Life: A Comprehensive Assessment of Water Management in Agriculture*. Earthscan and International Water Management Institute, London and Colombo.

Costanza, R., D'Arge, R., De Groot, R., et al. (1997). The value of the world's ecosystem services and natural capital. *Nature*, **387**, 253–260.

Crossman, N.D., Bryan, B.A. and Summers, N.D. (2011). Carbon payments and low cost conservation. *Conservation Biology*, **25**, 835–845.

Daily, G.C. (ed.) (1997). *Nature's Services: Societal Dependence on Natural Ecosystems*. Island Press, Washington, DC.

Daily, G.C., Polasky, S., Goldstein, J., et al. (2009). Ecosystem services in decision making: time to deliver. *Frontiers in Ecology and the Environment*, 7, 21–28.

DeFries, R. and Rosenzweig, C. (2010). Toward a whole-landscape approach for sustainable land use in the tropics. *Proceedings of the National Academy of Sciences*, **107**, 19627–19632.

de Groot, R.S. (1992). *Functions of Nature: Evaluation of Nature in Environmental Planning Management and Decision Making*. Wolters-Noordhoff, Amsterdam.

de Groot, R.S., Alkemade, R., Braat, L., Hein, L. and Willemen, L. (2010). Challenges in integrating the concept of ecosystem services and values in landscape planning, management and decision making. *Ecological Complexity*, 7, 260–272.

de Groot, R.S., Stuip, M.A.M., Finlayson, C.M. and Davidson, N. (2006). *Valuing Wetlands: Guidance for Valuing the Benefits Derived from Wetland Ecosystem Services*. Secretariat of the Convention on Wetlands and the Secretariat of the Convention on Biological Diversity, Gland and Montreal.

de Groot, R.S., Wilson, M.A. and Boumans, R.M.J. (2002). A typology for the classification, description and valuation of ecosystem functions, goods and services. *Ecological Economics*, **41**, 393–408.

Food and Agriculture Organization (2008). Scoping Agriculture–Wetland Interactions, Towards a Sustainable Multiple-Response Strategy. *FAO Water Reports*, **33**. Available at: http://www.fao.org/docrep/011/i0314e/i0314e00.htm (accessed August 2012).

Garrick, D., Siebentritt, M.A., Aylward, B., Bauer, C.J. and Purkey, A. (2009). Water Markets and freshwater ecosystem services: policy reform and implementation in the Columbia and Murray-Darling basins. *Ecological Economics*, **69**, 366–379.

Gordon, L.J., Finlayson, C.M. and Falkenmark, M. (2010). Managing water in agriculture for food production and other ecosystem services. *Agricultural Water Management*, **97**, 512–519.

Maltby, E. (2009). *Functional Assessment of Wetlands: Towards Evaluation of Ecosystem Services*. Woodhead Publishing, Cambridge.

Mander, Ü., Hayakawa, Y. and Kuusemets, V. (2005). Purification processes, ecological functions, planning and design of riparian buffer zones in agricultural watersheds. *Ecological Engineering*, **24**, 421–432.

Millennium Ecosystem Assessment (2003). *Ecosystems and Human Well-Being: A Framework for Assessment*. Island Press, Washington, DC.

Millennium Ecosystems Assessment (2005). *Ecosystems and Human Well-Being: Synthesis*. Island Press, Washington, DC.

National Research Council (2005). *Valuing Ecosystem Services. Toward Better Environmental Decision-Making*. National Academies Press, Washington, DC.

Perrings, C. (2006). Ecological economics after the millennium assessment. *International Journal of Ecological Economics and Statistics*, **6**, 8–22.

Porter, J., Costanza, R., Sandhu, H., Sigsgaard, L. and Wratten, S. (2009). The value of producing food, energy, and ecosystem services within an agro-ecosystem. *AMBIO: A Journal of the Human Environment*, **38**, 186–193.

Power, A.G. (2010). Ecosystem services and agriculture: tradeoffs and synergies. *Philosophical Transactions of the Royal Society B: Biological Sciences*, **365** (1554), 2959–2971.

Raudsepp-Hearne, C., Peterson, G.D. and Bennett, E.M. (2010). Ecosystem service bundles for analyzing tradeoffs in diverse landscapes. *Proceedings of the National Academy of Sciences*, **107**, 5242–5247.

Ribaudo, M., Greene, C., Hansen, L. and Hellerstein, D. (2010). Ecosystem services from agriculture: steps for expanding markets. *Ecological Economics*, **69**, 2085–2092.

Sandhu, H.S., Crossman, N.D. and Smith, F.P. (2012). Ecosystem services and Australian agricultural enterprises. *Ecological Economics*, **74**, 19–26.

Sandhu, H.S., Wratten, S.D. and Cullen, R. (2010a). Organic agriculture and ecosystem services. *Environmental Science and Policy*, **13**, 1–7.

Sandhu, H.S., Wratten, S.D. and Cullen, R. (2010b). The role of supporting ecosystem services in conventional and organic arable farmland. *Ecological Complexity*, **7**, 302–310.

Spangenberg, J.H. and Settele, J. (2010). Precisely incorrect? Monetising the value of ecosystem services. *Ecological Complexity*, **7**, 327–337.

Tallis, H., Kareiva, P., Marvier, M. and Chang, A. (2008). An ecosystem services framework to support both practical conservation and economic development. *Proceedings of the National Academy of Sciences*, **105**, 9457–9464.

TEEB (ed.) (2010). *The Economics of Ecosystems and Biodiversity: Ecological and Economic Foundations*. Earthscan, London.

Tscharntke, T., Klein, A.M., Kruess, A., Steffan-Dewenter, I. and Thies, C. (2005). Landscape perspectives on agricultural intensification and biodiversity – ecosystem service management. *Ecology Letters*, **8**, 857–874.

Turner, R.K., Georgiou, S. and Fisher, B. (2008). *Valuing Ecosystem Services: The Case of Multi-Functional Wetlands*. Earthscan, London.

Turner, R.K., van den Bergh, J.C.J.M., Söderqvist, T., et al. (2000). Ecological-economic analysis of wetlands: scientific integration for management and policy. *Ecological Economics*, **35**, 7–23.

Willemen, L., Hein, L. and Verburg, P.H. (2010). Evaluating the impact of regional development policies on future landscape services. *Ecological Economics*, **69**, 2244–2254.

Williams, W.D. (2001). Anthropogenic salinisation of inland waters. *Hydrobiologia*, **466**, 329–337.

Wilson, M.A. and Howarth, R.B. (2002). Discourse-based valuation of ecosystem services: establishing fair outcomes through group deliberation. *Ecological Economics*, **41**, 431–443.

Woodward, R.T. and Wui, Y.-S. (2001). The economic value of wetland services: a meta-analysis. *Ecological Economics*, **37**, 257–270.

Zedler, J.B. (2003). Wetlands at your service: reducing impacts of agriculture at the watershed scale. *Frontiers in Ecology and the Environment*, **1**, 65–72.

Zedler, J.B. and Kercher, S. (2005). Wetland resources: status, trends, ecosystem services and restorability. *Annual Review of Environment and Resources*, **30**, 39–74.

3

Key Ideas and Concepts from Economics for Understanding the Roles and Value of Ecosystem Services

Pamela Kaval[1] and Ramesh Baskaran[2]

[1] Havelock North, New Zealand and Marylhurst University, Oregon, USA
[2] Faculty of Commerce, Lincoln University, New Zealand

Abstract

Economists have been contributing to the discussion of the valuation of ecosystem services for many years; however, there is currently no standardization in the field. Consequently, studies differ extensively and comparisons between studies are difficult. This chapter briefly describes the primary economic methods commonly used to value ecosystem services. The results of an ecosystem service valuation literature review are then discussed. Finally, recommendations are offered on how to conduct ecosystem service valuation studies.

How can ecosystem services be valued?

It is easy to understand how ecosystem services contribute directly to life. For example, plants produce oxygen, a gas we need to breathe, while the ozone layer protects us from the sun's ultraviolet radiation. However, it is difficult to make comparisons between how much oxygen one tree produces, how much oxygen a person needs, how well the ozone layer prevents people from getting skin cancer, 50 tons of lumber, 3 hours of hiking and the 100 worms per square meter of soil that help to aerate the soil for plant growth. The easiest way to enable comparison of these ecosystem services is to use one type of unit. Economists have devised a methodology that enables us to use a dollar value as the common unit of comparison. Placing dollar values on ecosystem services makes it simpler for everyone, from farmers to politicians, to understand the value of a service,

Ecosystem Services in Agricultural and Urban Landscapes, First Edition. Edited by Steve Wratten, Harpinder Sandhu, Ross Cullen and Robert Costanza.

because most people use currency as a unit of value and medium of exchange (Costanza et al., 1997; Daily, 1997; Daily et al., 1997; de Groot et al., 2002).

Placing a dollar value on ecosystem services requires extensive reflection on the interconnectedness of ecosystems. As there are so many ecosystem services, there are also many ecosystem service values, from the price of gold, to the value of swimming in a stream, to the value of the safety of a fledgling in a bird nest on a tall cliff (Pearce and Turner, 1990; Merlo and Croitoru, 2005). By considering all ecosystem service values, the total economic value of nature is considered. The total economic value approach depends on the spatial and temporal scales being assessed, thus requiring analysts to be clear about the intended scope of their study. The total economic value conceptual framework views ecosystem goods and services as the flows of benefits and costs provided by the stock of natural capital (eftec, 2006).

Because there are so many types of ecosystem services, it is often preferable to group them together before attempting to calculate their value. The Millennium Ecosystem Assessment (2005) divided ecosystem services into four categories: supporting, provisioning, regulation and cultural services. Similarly, de Groot et al. (2002) also divided ecosystem services into four categories: regulation, habitat, production and information. In order to calculate the total economic value of ecosystem services, it may be easier to think of these services according to the type of value they provide (Fig. 3.1). Values can be assessed in the ways in which ecosystem services provide intangible benefits, or non-use values, where the resource is not directly used, and ways in which they support consumption, or use values, where the resource is being used.

More specifically, non-use values include altruistic, existence, bequest and option value. Altruistic value is the value people have knowing that others can enjoy goods and services from ecosystem services, even though they may never enjoy them themselves. For example, people may value knowing that others enjoy viewing the wildlife in Kenya's national parks and reserves, even though they will never go there to see the wildlife themselves. Existence value is derived from the satisfaction of knowing that a certain species or ecosystem exists, even if it will never be seen or used directly. An example of an existence value is knowing and feeling good about the existence of the blue whale, the largest mammal in the world living today. A person may believe that it is important that blue whales exist even though they may never see them although they may read about it in a book or see it on a television or movie programme. Bequest value is the satisfaction one obtains from being able to pass on environmental benefits to future generations. In this way, a person knows that the wildlife in Kenya's national parks and reserves will be available for their grandchildren and great grandchildren to visit someday. Option value pertains to the value people have knowing they have the option to use a resource in the future, even if they never do. This value relates to uncertainty and risk aversion, in that they are unsure they will ever use it, but don't want to risk the chance of it being lost.

Use values focus on the actual use of a resource and can be further subdivided into direct-use values, where a resource is directly being used in some way, and indirect-use values, where the resource is only indirectly being used. Direct use is further divided into extractive-use values that are extracted or consumed from

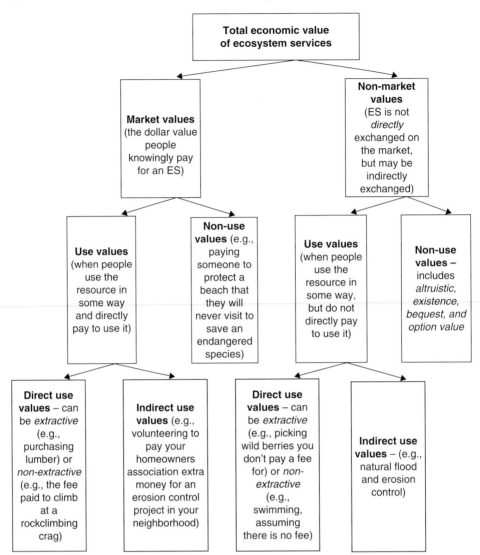

Fig. 3.1   Total economic value of ecosystem services. Note that market values are typically measured as direct use values; whereas indirect use and non-use values are more commonly measured as non-market values.

ecosystems, such as logging and fishing, and non-extractive-use values, from activities that are directly enjoyed, such as swimming, bird watching and cross-country skiing. The indirect-use value is referred to as a non-extractive-use value derived from functional services that the environment provides. For example, ecosystem regulatory processes that indirectly provide support and protection include erosion control and ultraviolet radiation protection (Freeman, 2003; National Research Council, 2005; Anderson, 2006; Tietenberg, 2006; Hanley and Barbier, 2009).

The next step is to determine whether the resource was paid for directly, as then it is considered a market value, or whether it was not paid for directly, or not paid for at all, as then it is a non-market value. For consumptive goods, when directly using a resource, such as eating a fish that you have purchased, we can consider the market value, in that a specific amount of money is exchanged in a market by people to directly use these products. When paying for something that will not be used directly, such as giving money to your neighbour's son to raise bees, you are experiencing an indirect-use market value. More specifically, since the son is raising bees for honey and not providing you with any of the honey, but you are still benefitting from the bees' pollination of the flowers in your yard, it is an indirect-use market value. However, if you donate money to sponsor a trip for your neighbour's son to work on an island to prevent poachers from stealing turtle eggs, you have a non-use market value because you feel good about saving the turtles even though you may never see them (Pearce and Turner, 1990; Freeman, 2003; National Research Council, 2005; Anderson, 2006).

Using similar examples, if you are fishing on your uncle's boat on the ocean and catch and eat a fish, but do not pay directly for this fish, as you do not need a fishing license to fish on the ocean, it is a non-market direct-use value. You have value in this trip, as you chose to go on the trip, may have paid for petrol to drive to your uncle's house and may pay to camp overnight somewhere to get there, but you did not pay 'directly' for the fish. If you did not give money to your neighbour's son for the bees, but the bees are still pollinating your flowers, you have an indirect non-market-use value. And if you did not give your neighbour's son any money to work on the island, but still feel good about him being there saving the turtles, you have a non-market non-use value for the turtles (Freeman, 2003; Anderson, 2006; Hanley and Barbier, 2009).

It is clear that a single person may benefit in more than one way from the same ecosystem. Thus, total economic value is the sum of all the relevant use and non-use, market and non-market values, for goods and services in a particular ecosystem. These measures of value can be included in policy and other land-management decisions.

Ecosystem service valuation methodologies

Economists have developed a number of market and non-market techniques to estimate the value of the environmental amenities from ecosystem services. Market values are calculated as out-of-pocket expenses and can be used to estimate the value of ecosystem goods and services that are traded in formal markets, such as the sale of timber and fish. Market values also include for example a decrease in the productivity of a fish stock, caused by an environmental effect such as an oil spill, that could lead to an earnings loss of a person dependant on fishing for their income. Defensive or preventive expenditures are another type of market value. These expenditures are made by a firm, government or individual to avoid or reduce an unwanted effect. An example of a defensive expenditure is the purchase of a water filter to drink water from a well contaminated by an unwanted chemical that leached into the groundwater system from a nearby mining operation.

Methods for measuring non-market values fall into two general categories: revealed preference and stated preference methods (Freeman, 2003; Hanley and Barbier, 2009). Revealed preference methods are based on observations of actual behaviour and allow us to make inferences about how individuals value changes in environmental quality. In contrast, stated preference measurements are based on responses to survey questions. Some common non-market valuation methods used today include the contingent valuation method, choice experiments, the travel cost method, and the hedonic pricing method. These methods are described briefly.

Revealed preference methods

The travel cost method sometimes called the Clawson Method, is a revealed preference method in that the respondent is revealing something that they actually did. Here, they report on the time they took and the costs they incurred to take a specific trip, costs that they would not have spent normally. An example is determining the cost of travelling to a lake to fish and camp. To do this, extra money is spent on fuel and camping fees, assuming the person already has all of their fishing equipment (Pearce and Turner, 1990; Haab and McConnell, 2002; Kahn, 2005; Anderson, 2006; Hackett, 2006).

Hedonic pricing is a revealed preference method that investigates the prices people pay for specific goods for the purpose of valuing an environmental resource. Oftentimes, the price that is investigated is a house/ property price. For example, to determine the value of seeing the beach from a house, the researcher could compare the price of houses overlooking a beach to equivalent homes one block away without a beach view (Hussen, 2000; Haab and McConnell, 2002; Kahn, 2005; Anderson, 2006; Hackett, 2006).

Stated preference methods

The contingent valuation method is sometimes called the willingness-to-pay or willingness-to-accept method. It is a stated preference method in that a person 'states' what they will do if a hypothetical situation were to arise. More specifically, they state how much they are willing-to-pay (willing-to-accept) for a change in a particular good or service. An example is the amount of money they would be willing-to-pay to hunt for deer in an area, if they were guaranteed to see at least some deer on a particular hunting trip (Hussen, 2000; Haab and McConnell, 2002; Daly and Farley, 2004; Kahn, 2005; Anderson, 2006; Hackett, 2006).

Choice modelling is a stated preference method in which a respondent is faced with a variety of alternatives and may be asked to select their most preferred alternative from a choice set (choice experiment), group their preferences (contingent grouping), rate their preferences (contingent rating), or rank their preferences (contingent ranking). There will typically be three or four alternative strategies with similar attributes (per question) presented to the respondents. An example of choice modelling alternatives include variations in the risk of toxic chemicals reaching the groundwater, the percentage of harvested trees, the percentage of species diversity, as well as a dollar value, such as an entrance fee or a fee in your annual taxes/ rates (Louviere et al., 2000; Bateman et al., 2002; Hensher et al., 2005; Street and Burgess, 2007; Riera et al., 2012).

These four methods, together with direct-market values, can aid us in valuing many ecosystem services. But they fall short of valuing all ecosystem services, for which other methods must be employed. These include the avoided cost method, the replacement cost method, the restoration cost method, factor income, and the benefit transfer method.[1]

Other methods

Avoided cost methods   attempt to quantify the costs we do not have to pay when nature is providing a particular good. One example is to calculate the value of storm and buffer functions provided by coastal wetlands in the event of a hurricane or cyclone. To do this, you could calculate the potential financial losses if the wetlands did not exist. In 2005, Hurricane Katrina caused over $US81 billion damage to the New Orleans area. If the wetlands around New Orleans had not been destroyed by years of alterations to the Mississippi River, New Orleans would not have been almost completely exposed to the Gulf of Mexico, and there may not have been any, or as much, damage (Daily, 1997; Daily, 1997; Knabb, 2006; Cleveland, 2006).

Replacement cost   is a method used to calculate the cost of replacing a service with a human-created product, such as fertilizers to replace the nutrients that are recycled by earthworms and benefit the soil (Hussen, 2000; Kahn, 2005).

Restoration cost   is a method used to calculate the cost of restoring an ecosystem to the natural state that existed prior to an environmental damage, such as the cost of repairing the environmental damage caused by the Exxon Valdez Oil Spill of 1989 (Bragg et al., 1994; Kahn, 2005).

Factor income   is the value of an ecosystem service that enhances the market value of ecosystem services. For example, bees pollinate the flowers of the agricultural crops sold on the market (Woodward and Wui, 2001; Brander et al., 2006). The marginal benefit of pollination services to the crop can be used to estimate the value of the service provided by the bees.

---

[1]Some studies also consider group valuation or discourse based methods to obtain values for ecosystem services. In a discourse based study, people get together in a designated location and discuss their values for an ecosystem good or service. Since ecosystem services are commonly public goods that affect many people, some feel that the valuation of these public services should not come from individual-based values, such as in the previous approaches used, but from public discussion. In this way, the values derived are considered those of society and are believed to lead to socially equitable and politically legitimate outcomes (Wilson and Howarth, 2002). Consequently, this method focuses on qualitative values. The focus of this study is quantitative methods, therefore, this method is not being considered here.

Benefit transfer   or value transfer, is a method used as a result of time limitations and/or budget constraints and focuses on applying secondary data. In this
method, a researcher uses existing economic valuation information from a study
conducted in a particular area, called the study site, and transfers those values to
a new site or area, sometimes called the policy site. Care should be made to transfer values from an area that is similar to the policy site (Kaval and Loomis, 2003;
Kahn, 2005). There are two types of benefit transfers: value transfers and
function transfers. A value-transfer approach takes a single point estimate, usually
a mean willingness-to-pay or an average of point estimates from multiple studies
that have been developed elsewhere, to transfer to a new study area. A function
transfer approach transfers the entire estimated equation (function) of a study
site to the policy site. For example, a travel cost demand equation from a study
site could be used with the socioeconomic or demographic characteristics such as
income, average travel costs and quality conditions at the policy site to estimate
the average willingness-to-pay of different proposed plans at the policy site.
While this method is listed under non-market valuation methods, it can also be
used to transfer market values.

   Table 3.1 is an extension of the de Groot et al. (2002) table and provides a list
of ecosystem services, their value types, as well as the methods commonly used
to calculate their dollar value. As can be seen, researchers use different methods
to calculate values. Recreation, for example, is a direct use value and can be calculated as a market or non-market value. If you paid money to use an indoor
climbing wall, the price paid is a market value. However, if you went to climb in
a park that does not charge an entrance fee, this would be considered a non-
market value. Non-market-valuation methods commonly used to calculate recreation values include the contingent-valuation method, travel-cost method,
choice experiments, factor income, hedonic method, avoided costs, restoration
costs and the benefit-transfer method. Science and education, on the other hand,
are considered a market value and a direct use. Valuation methods commonly
used for science and education include market valuation and benefit transfer
(Hartwick and Olewiler, 1998; de Groot et al., 2002; Kahn, 2005).

   As can be seen, the valuation method used will depend on the type of service
being studied. Many different methods can work for any given service, and the
method of choice depends on the availability of the resources, time, data, specific
characteristics and goals of the study.

How ecosystem services have been measured in the past

Ecosystem service studies are well represented in the literature, even if they were
not always termed as such. One of the first and most thorough, original longitu-
dinal ecosystem service studies that predated this discipline was a Rhone Poulenc
farm management study conducted by Higgenbotham et al. (Higginbotham et al.,
1997, 1999, 2000). In this seminal study that began in 1994 on 57 hectares
in Essex, they compared organic farming to reduced input and conventional
farming for a variety of crops. They not only estimated the values, costs and
yields of the crops, but also measured food quality, the taste of the final goods,

Table 3.1 Ecosystem services and the commonly used methods for calculating their dollar values. This table is based on data from de Groot et al. (2002), Table 2, with modification and extension. It includes commonly used valuation techniques, but is not exhaustive.

Ecosystem service	Market or non-market good	Use or non-use values	Valuation methods
1 Science and education	Market	Direct use	Market valuation, benefit transfer
2 Recreation	Market and non-market	Direct use	Market valuation, contingent valuation, travel cost, choice experiments, factor income, hedonic pricing, avoided costs, restoration costs, benefit transfer
3 Genetic and medicinal resources	Market and non-market	Direct use and indirect use	Market valuation, factor income, benefit transfer
4 Raw materials	Market and non-market	Direct use and indirect use	Market valuation, factor income, contingent valuation, choice experiments, benefit transfer
5 Food production	Market and non-market	Direct use and indirect use	Market valuation, factor income, contingent valuation, choice experiments, benefit transfer
6 Nursery function	Market and non-market	Direct use and indirect use	Market valuation, contingent valuation, avoided costs, replacement cost, factor income, choice experiments, restoration costs, benefit transfer
7 Plant and animal refugia	Market and non-market	Direct use, indirect use and non-use	Market valuation, contingent valuation, choice experiments, restoration costs, benefit transfer
8 Soil formation	Market and non-market	Direct use, indirect use and non-use	Market valuation, avoided costs, benefit transfer
9 Purification and regulation of air and water	Market and non-market	Indirect use	Market valuation, avoided costs, replacement cost, factor income, contingent valuation, choice experiments, benefit transfer

(continued)

Table 3.1 *(cont'd)*

Ecosystem service	Market or non-market good	Use or non-use values	Valuation methods
10 Natural pests and biological control	Market and non-market	Indirect use	Market valuation, replacement cost, factor income, restoration costs, benefit transfer
11 Detoxification and decomposition of wastes	Non-market	Indirect use	Contingent valuation, replacement costs, choice experiments, benefit transfer
12 Protection for the sun's ultraviolet rays	Non-market	Indirect use	Contingent valuation, replacement cost, choice experiments, restoration costs, benefit transfer
13 Partial climate stabilization	Non-market	Indirect use	Avoided cost, benefit transfer
14 Natural disturbance regulation	Non-market	Indirect use	Avoided cost, replacement cost, benefit transfer
15 Erosion control	Non-market	Indirect use	Avoided cost, replacement cost, restoration cost, benefit transfer
16 Plant pollination	Non-market	Indirect use and non-use	Avoided cost, replacement cost, factor income, benefit transfer
17 Nutrient recycling	Non-market	Indirect use and non-use	Replacement cost, benefit transfer
18 Seed dispersal	Non-market	Non-use	Avoided cost, replacement cost, benefit transfer
19 Biodiversity maintenance	Non-market	Non-use	Contingent valuation, choice experiments, restoration costs, avoided costs, benefit transfer
20 Aesthetic beauty	Non-market	Non-use	Contingent valuation, choice experiments, benefit transfer
21 Human culture	Non-market	Non-use	Contingent valuation, choice experiments, benefit transfer
22 Preservation (including existence, bequest and option value)	Non-market	Non-use	Contingent valuation, choice experiments, benefit transfer

earthworm populations, weed populations and insect populations. This project began in the early 1980s and is believed to be still in progress today (crop research at http://www.hgca.com).

As the discipline has advanced, the term ecosystem service valuation has been used more often. In conducting this investigation, it was discovered that some researchers that conduct an ecosystem service valuation study focused on one particular value (such as recreation), some focused on one particular valuation method (typically contingent valuation or choice experiments), some solely present conceptual models, while others focus on using a benefit-transfer or value-transfer approach, such as the Constanza et al. (1997) study. Several researchers have used the values presented in the Costanza et al. (1997) work to create their own estimates, which is technically a benefit transfer of a benefit transfer. All of these works provide some insight into ecosystem service valuation. However, this investigation focused on the articles that attempted to value three or more ecosystem services using original data and more than one valuation method, of which there was disappointingly only a select few (Table 3.2). Due to practical limitations, the studies listed here may not be exhaustive, although strenuous attempts were made to include all studies that fit the guidelines.

As can been seen in Table 3.2, a variety of systems have been investigated, from alpine areas to coral reefs. In relation to valuation, all studies included a market valuation aspect, but that is where the similarities between the studies dissipate. Some focus on obtaining information by surveying respondents, while others focus on using field data to calculate values. Except for the two studies by Cesar et al., the type of ecosystem services studied also varied extensively. Overall, the study by Sandhu (2007) is believed to be the most comprehensive of all ecosystem services studies to date using original data (Johnston et al., 2002; Cesar and Van Beukering, 2004; Cesar et al., 2004; Sandhu, 2007; Grêt-Regamey et al., 2008; Jenkins et al., 2010).

Ecosystem service valuation study recommendations

As demonstrated in Tables 3.1 and 3.2, there are no standard methodologies for ecosystem service valuation. Therefore, when planning to conduct an ecosystem services valuation, it is important to consider the strengths and weaknesses of candidate methods and how these are related to your research objectives. For example, if you are investigating grazing impacts on the Sevilleta National Wildlife Refuge, a 93 000-ha state park located in New Mexico (USA), you should firstly determine the ecosystem types that may be impacted. This particular park represents a variety of ecosystem types, including the Chihuahuan Desert, Great Plains Grassland, Great Basin Shrub-Steppe, Piñon-Juniper Woodland, Bosque Riparian Forests, Wetlands and Montane Coniferous Forest (United States Fish and Wildlife Service, 2007). Grazing cattle on the Great Plains Grassland area may have impacts on adjacent ecosystems and therefore extend the study area. You might also want to consider the broader effects, such as clean air provision through photosynthesis, as well as effects on downstream aquatic values.

Table 3.2 Summary of original ecosystem service valuation studies valuing three or more ecosystem services focusing on a particular area and using more than one valuation method (Johnston et al., 2002; Cesar and Van Beukering, 2004; Cesar et al., 2004; Sandhu, 2007; Grêt-Regamey et al., 2008; Jenkins et al., 2010).

Author(s)	Year	System	Location	Ecosystem services studied	Valuation methods
Cesar et al.	2004	Coral reefs	Hawaii, USA	Recreation, fishery, amenity, biodiversity research	Market values, travel cost, replacement cost, contingent valuation
Cesar et al.	2004	Marine ecosystem	Seychelles, Republic of Seychelles	Recreation, fishery, amenity, biodiversity research	Market values, travel cost, contingent valuation
Grêt-Regamey et al.	2008	Alpine	Swiss Alps, Europe	Avalanche protection, timber production, scenic beauty, habitat	Market values, contingent valuation, replacement cost, GIS measurements converted to market values
Jenkins et al.	2010	Forested wetland	Mississippi alluvial valley, USA	Greenhouse gas mitigation, nitrogen mitigation, waterfowl recreation	Market values, field measurements converted to market and social values, avoided costs, benefit transfer
Johnston et al.	2002	Estuary	New York, USA	Amenity, recreation, nursery services, habitat refugia, preservation	Market values, choice experiment, travel cost, hedonic method
Sandhu	2007	Arable farmland	Canterbury, New Zealand	Food, hydrological flow, aesthetics, raw materials, recreation, fuel wood, science and education, conservation of species, maintenance of genetic resources, pollination, mineralization of plant nutrients, support to plants, biological pest control, soil fertility, soil formation, carbon accumulation, soil erosion, nitrogen fixation, and services provided by shelterbelts	Market prices, field measurements converted to market values, avoided costs, GIS measurements converted to market values

There are many issues that can be considered in an ecosystem service valuation study, but given the practical constraints on time and funding, the following guidelines should be helpful in focusing your investigation:

1 Define the boundaries of your research area (e.g. all of Sevilleta National Wildlife Refuge, all areas the refuge effects or only the Great Plains Grassland area).
2 Define the ecosystem types located in your research area (i.e. in the Sevilleta National Wildlife Refuge, it may include the Chihuahuan Desert, Great Plains Grassland, Great Basin Shrub-Steppe, Piñon-Juniper Woodland, Bosque Riparian Forests, Wetlands and Montane Coniferous Forest).
3 Determine what ecosystem services (Table 3.1) are currently functioning (or could be functioning if something is changed) in the research area.
   a Determine which people benefit from these services (location, demographics).
   b Determine the scarcity of these services in the region.
   c Determine whether these services have readily available regional natural or man-made substitutes.
   d Determine whether these services are restorable in this area.
4 Determine the ecosystem response to the changes being investigated (e.g. an invasive species enters the area, grazing stops, grazing begins, the land is paved over and no ecosystem services exist there anymore).
   a Determine which of the defined ecosystem services will change.
   b Determine different scenarios for the types of change possible (e.g. the invasive species spreading quickly vs. the invasive species spreading slowly).
5 Determine whether these services of interest are market or non-market use or non-use values.
6 Determine the most appropriate valuation methods to use to value the ecosystem services, given your objectives, as well as your funding and time constraints.
7 Conduct your research according to the guidelines you have defined.

Conclusions

According to the Millennium Ecosystem Assessment (2005), over the last half of the twentieth century, humans have been rapidly and extensively affecting ecosystems and their services, resulting in substantial and irreversible biodiversity losses, while attempting to meet world-wide demands for consumption of goods and services. A recent case in point is the 2010 Gulf of Mexico BP/ Deepwater Horizon Oil Spill, one of the largest offshore oil spills in history. This oil spill covered over 2500 square miles (6500 km^2) in area. This spill has caused, and will continue to cause, extensive damage for decades to come, not only to human activities of fishing and tourism, but also to the plant and animal species. If we were to place a value on this spill and only consider the losses to the fishery, and the tourism industry, we would be making a significant underestimate of the ecosystem service value of this great water resource (McDonald et al., 2006). Economists recognize there are multiple types of direct and indirect values associated with ecosystems. A range of market and non-market valuation methods have been developed over the last 40 years to provide estimates of ecosystem service values including Factor Income, Travel Cost, Contingent Valuation and Choice Modelling. Valuation of ecosystem services needs considerable care to ensure that reliable and valid estimates are provided to policy makers.

References

Anderson, D.A. (2006). *Environmental Economics and Natural Resource Management*, 2nd edn. Pensive Press. Danville, Kentucky.

Bateman, I., Carson, R.T., Day, B., et al. (2002). *Economic Valuation with Stated Preference Techniques: A Manual*. Edward Elgar, Cheltenham.

Bragg, J.R., Prince, R.C., Harner, E.J. and Atlas, R.M. (1994). Effectiveness of bioremediation for the Exxon Valdez oil spill. *Nature*, 368, 413–418.

Brander, L., Florax, R. and Vermaat, J. (2006). The empirics of wetland valuation: a comprehensive summary and a meta-analysis of the literature. *Environmental and Resource Economics*, 33, 223–250.

Cesar, H.S.J. and Van Beukering, P.J.H. (2004). Economic valuation of the coral reefs of Hawai'i. *Pacific Science*, 58, 231–242.

Cesar, H.S.J., van Beukering, P.J.H., Payet, R. and Grandourt, E. (2004). *Evaluation of the socio-economic impacts of marine ecosystem degradation in the Seychelles*. Cesar Environmental Economics Consulting, the Netherlands.

Cleveland, C.J., Betke, M., Federico, P., et al. (2006). Economic value of the pest control service provided by Brazilian free-tailed bats in south-central Texas. *Frontiers in Ecology and the Environment*, 4, 238–243.

Costanza, R., dArge, R., deGroot, R., et al. (1997). The value of the world's ecosystem services and natural capital. *Nature*, 387, 253–260.

Daily, G.C. (1997). *Nature's Services: Societal Dependence on Natural Ecosystems*. Island Press, Washington, DC.

Daily, G.C., Alexander, S., Ehrlich, P.R., et al. (1997). Ecosystem services: benefits supplied to human societies by natural ecosystems. *Issues in Ecology*, 2, 1–16.

Daly, H.E. and Farley, J. (2004). *Ecological Economics: Principles and Applications*. Island Press, Washington, DC.

de Groot, R.S., Wilson, M.A. and Boumans, R.M.J. (2002). A typology for the classification, description and valuation of ecosystem functions, goods and services. *Ecological Economics*, 41, 393–408.

eftec (2006). *Valuing our Natural Environment*. eftec, London.

Freeman, A.M. (2003). *The Measurement of Environmental and Resource Values Resources for the Future*. RFF Press, Washington D.C.

Grêt-Regamey, A., Bebi, P., Bishop, I.D. and Schmid, W.A. (2008). Linking GIS-based models to value ecosystem services in an Alpine region. *Journal of Environmental Management*, 89, 197–208.

Haab, T.C. and McConnell, K.E. (2002). *Valuing Environmental and Natural Resources: The Econometrics of Non-Market Valuation*. Edward Elgar, Cheltenham.

Hackett, S.C. (2006). *Environmental and Natural Resource Economics: Theory, Policy, and the Sustainable Society*, 3rd edn. M.E. Sharpe, New York.

Hanley, N. and Barbier, E.B. (2009). *Pricing Nature. Cost–Benefit Analysis and Environmental Policy*. Edward Elgar.

Hartwick, J.M. and Olewiler, N.D. (1998). *The Economics of Natural Resource Use*, 2nd edn. Addison-Wesley, Boston, Massachusetts.

Hensher, D.A., Rose, J.M. and Greene, W.H. (2005). *Applied Choice Analysis: A Primer.* Cambridge University Press, New York.

Higginbotham, S., Noble, L., Beck, R. and Brown, R. (1999). *Rhone-Poulenc Agriculture, a Division of Aventis Crop Science UK Limited: Farm Management Study.* Aventis Crop Science, Ongar Essex, England.

Higginbotham, S., Noble, L., Beck, R. and Brown, R. (2000). *Rhone-Poulenc Agriculture, a Division of Aventis Crop Science UK Limited: Farm Management Study.* Aventis Crop Science, Ongar Essex, England.

Higginbotham, S., Noble, L., Turner, R., et al. (1997). *Rhone-Poulenc Agriculture: Farm Management Study.* Rhone-Poulenc Agriculture, Ongar Essex, England.

Hussen, A. (2000). *Principles of Environmental Economics*, 2nd edn. Routledge, New York.

Jenkins, W.A., Murray, B.C., Kramer, R.A. and Faulkner, S.P. (2010). Valuing ecosystem services from wetlands restoration in the Mississippi alluvial valley. *Ecological Economics*, **69**, 1051–1061.

Johnston, R.J., Grigalunas, T.A., Opaluch, J.J., Mazzotta, M. and Diamantedes, J. (2002). Valuing estuarine resource services using economic and ecological models: the Peconic estuary system study. *Coastal Management*, **30**, 47–65.

Kahn, J.R. (2005). *The Economic Approach to Environmental and Natural Resources*, 3rd edn. Thomson South-Western, Independence, Kentucky.

Kaval, P. and Loomis, J. (2003). *Updated Outdoor Recreation Use Values with Emphasis on National Park Recreation. A Report Prepared for the National Park Service.*

Knabb, R.D., Rhome, J.R. and Brown, D.P. (2006). *Tropical Cyclone Report: Hurricane Katrina: 23–30 August (2005).* National Hurricane Center, Miami, Florida.

Louviere, J.J., Hensher, D.A., Swait, J.D. and Adamowicz, W. (2000). *State Choice Methods: Analysis and Applications.* Cambridge University Press, New York.

McDonald, S., Oates, C.J., Young, C.W. and Hwang, K. (2006). Toward sustainable consumption: Researching voluntary simplifiers. *Psychology and Marketing*, **23**, 515–534.

Merlo, M. and Croitoru, L. (2005). *Valuing Mediterranean Forests: Towards Total Economic Value.* CABI Publishing, Oxford, UK.

Millennium Ecosystem Assessment (2005). *Ecosystems and Human Well-Being: Synthesis.* Island Press, Washington, DC.

National Research Council (2005). *Valuing Ecosystem Services: Toward Better Environmental Decision-Making.* National Academies Press, Washington, DC.

Pearce, D.W. and Turner, R.K. (1990). *Economics of Natural Resources and the Environment.* Pearson Education Limited, Essex.

Riera, P., Signorello, G. and Thiene, M. et al. (2012). Non-market valuation of forest goods and services: Good practice guidelines. *Journal of Forest Economics*, in press.

Sandhu, H.S. (2007). *Quantifying the Economic Value of Ecosystem Services on Arable Farmland: A Bottom Up Approach.* Lincoln University, Christchurch, New Zealand.

Street, D.J. and Burgess, L. (2007). *The Construction of Optimal State Choice Experiments: Theory and Methods.* Wiley-Interscience, Hoboken, New Jersey.

Tietenberg, T. (2006). *Environmental and Natural Resource Economics*, 7th edn. Pearson, Boston, Massachusetts.

United States Fish and Wildlife Service (2007). *Sevilleta National Wildlife Refuge.* Available at: http://www.fws.gov/southwest/REFUGES/newmex/sevilleta/ (accessed January 14, 2010).

Woodward, R.T. and Wui, Y.S. (2001). The economic value of wetland services: a meta-analysis. *Ecological Economics*, **37**, 257–270.

Part B

Ecosystem Services in Three Settings

4

Viticulture can be Modified to Provide Multiple Ecosystem Services

Sofia Orre-Gordon,[1,2] Marco Jacometti,[2] Jean Tompkins[2] and Steve Wratten[2]

[1] Barbara Hardy Institute, University of South Australia, Adelaide, Australia
[2] Bio-Protection Research Centre, Lincoln University, New Zealand

Abstract

In wine-producing regions, such as California and New Zealand, the wine industry has an increasing importance to national economies and growers have responded by increasing the area of land devoted to this crop. These virtual monocultures depend on high agrochemical input to control pest, disease and weed problems. This chapter covers different habitat modification methods that can be deployed to enhance existing naturally occurring ecosystem services within vineyards in an attempt to reduce the reliance on synthetic chemicals and increase the sustainability of the wine production.

Introduction

Little did the ancient Greeks know that one day the vine, grown for the wine that symbolized Dionysus, would occupy a vast bleak landscape, with bare soil or short mowed ryegrass on the ground between immeasurable rows of *Vitis vinifera* L. (Vitaceae), as is the case today.

Currently, around 2.7 million tonnes of wine is produced each year, primarily by Italy, France and the United States (FAO, 2011). Wine production in these and other wine-producing regions is rapidly expanding in area and volume, raising the importance of this crop to their respective national economies. In California, the fourth largest wine producer in the world, the wine industry has an annual impact of \$US51.8 billion on the state's economy and an impact of \$US125.3

Ecosystem Services in Agricultural and Urban Landscapes, First Edition. Edited by Steve Wratten, Harpinder Sandhu, Ross Cullen and Robert Costanza.
© 2013 John Wiley & Sons, Ltd. Published 2013 by John Wiley & Sons, Ltd.

billion on the US economy as a whole. The industry creates 875 000 jobs throughout the USA and wine industry-related employment has increased by 37% since 2002, despite an increasingly competitive market environment (Wine Institute, 2006). Similar trends may be seen in New Zealand where wine production is one of the country's top export earners. In the year 2008, the value of wine exports increased by 24% (Beef + Lamb New Zealand, 2010) making the wine industry worth close to $NZ1 billion for this South Pacific nation. Stimulated by this success, the area of wine grape production is rapidly rising with a 71% increase over the last 10 years, resulting in a total production area of 31 057 ha in 2009 (Aitken and Hawlett, 2010). Clearly, wine grapes are increasing in importance to national economies and growers have responded by increasing the area of land devoted to this crop.

This chapter covers different habitat modification methods that can be deployed to enhance existing, naturally occurring ecosystem services. Consequently, it will not cover inundative biological control methods, such as the application of 'biofungicides' and 'biopesticides' where pest control is achieved by the released biocontrol agents themselves and is often aimed at a shorter time period.

Rather it will look at how vineyards can be modified to enhance ecosystem services (see Chapter 1), especially that of conservation biological control (CBC) – a form of habitat manipulation to improve pest management. Enhancing CBC in vineyards has the potential to improve wine production sustainability as it can reduce the reliance growers have on external synthetic pesticide inputs.

Enhancing CBC in vineyards

Typically, vineyards are virtual monocultures which depend on high agrochemical input to control pest, disease and weed problems. This reliance on synthetic chemicals is high, partially because the simplified vineyard environment is inhospitable to natural enemies of pests, who consequently do not inhabit the vineyard in numbers sufficient to effectively control pest populations. Pesticides have come to replace natural enemies in the role of pest control. More recently, however, with the increased awareness of the need to find alternatives to pesticides, methods are being developed to bring back natural enemies to the vineyard and consequently reinstate a much-valued ecosystem service (Gurr et al., 2007).

A lot of work on habitat manipulations utilized to gain ecosystem function enhancement has been done within CBC (Jonsson et al., 2010). In CBC, habitat manipulation techniques are used to produce trophic cascades. These result in inverse patterns of abundance or biomass across more than one trophic level. In a three-trophic-level food chain, such as crop plants–herbivorous pests–natural enemies, enhancing the top predators (natural enemies) may result in lower abundance of mid-level consumers (herbivores pest) and a higher abundance of basal producers (crop plants) (Carpenter and Kitchell, 1993).

CBC utilizes these 'top-down' effects to increase the natural enemy population (Gurr et al., 2000). However, habitat manipulation can produce both 'bottom-up'

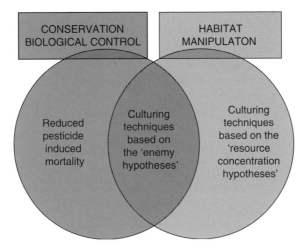

Fig. 4.1 Conservation biological control shares common techniques with habitat manipulation but are not synonymous. (Data from Gurr et al., 2000).

and 'top-down' effects, consistent with the 'resource concentration' hypothesis and the 'enemy' hypothesis. According to both these hypotheses, herbivores are predicted to be more abundant in simple systems, that is monocultures, than in more complex ones (Root, 1973). According to the 'resource concentration' hypothesis, the reduction in herbivore abundance in complex habitats is caused by mechanisms such as dilution of the contrast between a concentrated crop and the soil. This produces a dilution of the visual and chemical stimuli for the herbivore, resulting in decreased colonization rates and increased emigration rates and thereby a reduction in damage to the crop (Gurr et al., 2000). As the herbivore population in the 'resource concentration' hypothesis is determined by a lower trophic level, the effects seen are 'bottom-up' effects. According to the 'enemy' hypothesis, predators and parasitoids are more numerous and/or effective in more diverse systems than in simple ones (Root, 1973). As the herbivore population in the 'enemy' hypothesis is impacted by a higher trophic level, the effects seen are 'top-down' effects. These effects are utilized in CBC, which specifically involves maximization of the impact of natural enemies by providing key ecological resources and by minimizing pesticide-induced mortality (Gurr et al., 2000) (Fig. 4.1).

In vineyards the main focus of habitat manipulation work has been on the control of leafrollers and *Botrytis cinerea* (Helotiales: Sclerotiniaceae). Therefore, the main emphases of this chapter will be given to these two problems and the management of them. Several other global pest and diseases commonly cause problems in vineyards, such as mealy bugs (Signoret) (*Planococcus ficus* Hemiptera: Pseudococcidae), Japanese beetles (Newman) (*Popillia japonica* Coleoptera: Scatabaeidae), mildew, black rot and leaf spot. However, no approaches on enhancing naturally occurring ecosystem services have been used to control them. Alternative habitat manipulation techniques than the local environmental manipulations discussed here to control *B. cinerea* have recently been reviewed by Jacometti et al. (2010).

Leafrollers and *Botrytis cinerea* in the vineyards

Two of the main yield-reducing problems in vineyards are leafrollers, especially the highly polyphagous light brown apple moth, *Epiphyas postvittana* (Walker) (Lepidoptera: Tortricidae) and the fungus, *B. cinerea*. The leafroller larvae feed on new shoots, flowers, stalks and leaves.

Leafroller management methods in many vineyards involve the use of synthetic broad-spectrum insecticides. These are applied on a calendar basis with little regard to pest abundance (Gurr et al., 2007). In California, about 10 million kg of active ingredients of pesticides are used annually to control pest pressures in wine grapes. Despite an overall decrease in all pesticide usage across commodities, in wine grapes there has been a 3% increase in active ingredients usage over the last year (Schwarzenegger et al., 2010).

Leafroller larvae not only damage the grapevines but are one of the factors making the grapes more susceptible to infection by botrytis, causing bunch rot in the damaged bunches. Bunch rot in New Zealand may cause midseason losses exceeding 20% and in very wet seasons may cause complete crop losses. The fungus can also affect flowers, leaves, buds, shoots, stems and/or fruits, often limiting plant development, fruit-set, yield and fruit quality. Botrytis is most commonly managed through canopy pruning and the prophylactic use of fungicides (Jacometti et al., 2010).

Habitat modification to enhance naturally occurring pest control

Both the development of resistance among the pest organisms and public concerns of the effects of the synthetic chemicals on human health and the environment have made high agrichemical input management strategies undesirable. Consequently, alternative methods to control the problem species have been developed. These often rely on the deployment of habitat modification methods to enhance the existing controlling mechanism.

Floral resource supplementation as a form of habitat modification

One way to potentially enhance the biodiversity within a vineyard and, at the same time, increase biological control of pest species is through the application of CBC. In CBC, natural enemies of pests are provided with floral resources. Growing non-crop plants, such as flowering plants within or around the crop from which natural enemies can benefit, may enhance their controlling efficiency of the pest (Tylianakis et al., 2004). CBC is based on Root's 'enemies' hypothesis where natural enemies are more abundant in diverse crop environments (Root, 1973). This implies that habitat management in the form of increased diversification can be used to conserve and enhance natural enemies (Jonsson et al., 2010).

In CBC, omnivorous natural enemies are provided with alternative food sources, such as nectar and/or pollen. This may prevent them from starving or emigration while prey/ host densities are temporarily low in the area. The floral supplementation may also increase longevity, fecundity and other components of 'ecological fitness', which may in turn increase the pest controlling efficiency of the natural enemies.

One way to reduce the damage by leafrollers is by increasing the abundance of parasitoids and predators attacking the herbivore through habitat modification in the form of floral resource supplementation. These resources are aimed to increase the 'ecological fitness' of natural enemies and subsequently, hopefully lead to increased biological control of the pest through top-down mechanisms.

In New Zealand, the most common natural enemy attacking leafroller larvae is an endoparisitic braconid parasitic wasp, *Dolichogenidea tasmanica* (Cameron) (Hymenoptera: Braconidae). This parasitoid also controls the leafroller population in Australia together with the egg parasitoid *Trichogramma carver* (Carver) (Hymenoptera: Trichogrammatidae), the techinid fly *Voriella uniseta* (Malloch) (Diptera: Tachinidae), and the larva of lacewings such as *Micromus tasmaniae* (Walker) (Neuroptera: Hemerobiidae) (Gurr et al., 2007).

Floral resources commonly used as a habitat manipulation tool is buckwheat *Fagopyrum esculentum* (Moench) (Ploygonaceae) and alyssum *Lobularia maritima* (L.) (Brassicaceae). These have nectar and pollen that is easily accessible by natural enemies of herbivores and tend not to increase the fitness of the pest or the natural enemies of the natural enemies. Under laboratory conditions, *D. tasmanica* with access to flowering buckwheat (Scarratt, 2005) and alyssum (Berndt and Wratten, 2005) has had an increased fecundity and longevity and the proportion of female offspring increased. The maximum longevity of *D. tasmanica* increased seven fold from 2.2 ± 0.17 days to 15.7 ± 2.77 days when it had access to alyssum flowers. As a result of the increase in longevity the lifetime fecundity increased by almost eightfold. Also, the sex ratio of the offspring of the parasitoids changed from strongly male biased with a mean sex ratio close to 1 to around 0.6 when alyssum flowers where present (Berndt and Wratten, 2005). Adding alyssum to a vineyard scenario has been shown, in some instances, to increase parasitism rates of leafrollers close to the flowering plants and to decrease the leafroller densities (Scarratt, 2005). Also, flowering buckwheat has been successful in increasing the parasitism rate of leafrollers in vineyards and has, in some instances, increased the rate by more than 50% (Berndt et al., 2006). Provision of flowering buckwheat has now been adopted as a measure to control the herbivore in vineyards in all the major wine regions in New Zealand and Australia (Fig. 4.2).

Mulch application as a form of habitat modification

There are several alternatives to synthetic fungicides for *B. cinerea* management in vineyards. These have been reviewed by Jacometti et al. (2010). One way to modify the vineyard environment to provide increased biological control of *B. cinerea* is through the enhancement of soil microbial activity. The

Fig. 4.2 Buckwheat, *Fagopyrumesculentum* planted between vine rows in a New Zealand vineyard to enhance biological control of leafrollers.

microbial activity can be increased by adding organic (plant based) mulches and cover crop mulches in situ. Such habitat modification can be highly effective in managing *B. cinerea* (Jacometti et al., 2007a,b,c). Mulches can reduce *B. cinerea* primary inoculum through increased competition from elevated soil biota in response to the mulch application and increased degradation of the residual vine material that functions as a host for the fungus (Jacometti et al., 2007b).

Mulch application can reduce the conidiophore coverage on vine debris by 66–95% compared to no mulch application (Jacometti et al., 2007b). Fermented or composted grape *marc* (grape pressings) and shredded office paper have been shown to increased soil biological activity by two to four times (Kratz, 1998; Girvan et al., 2003).

The increased soil biological activity due to the mulch application can also lead to changes in soil attributes such as increased availability of soil nutrients and water for the plant, which may increase the vines' resistance to the disease. The resistance can be due to decreased vine canopy density and/or increased grape skin strength. Jacometti et al. (2007c) showed a decreased by up to 30% in canopy density in response to mulch application. This may cause a reduction in canopy humidity and increased light penetration, canopy temperature and photosynthetic rate. The change in canopy density alone or in combination with the changes in the above mentioned soil attributes can increase grape skin strength by up to 10% in paper mulch treatments. These changes to the soil and vine environment have been shown to reduce botrytis bunch rot averaged up to 97% over two consecutive harvests compared to the control treatments with no marc and paper added. Under high disease pressure adding paper and marc to Reisling grapes in New Zealand has shown to reduce *B. cinerea* severity on the grapes to below the economic threshold (Jacometti et al., 2007c).

Combining two forms of habitat modification

As mentioned above, the application of mulches to the vineyard habitat can be a successful method to control *B. cinerea*. Using plants suitable for resource supplementation for natural enemies such as *Phacelia* (Benth) (*Phacelia tanacetifolia* cv. Balo, Boraginaceae) as mulch in the under-vine area could combine the two biological control functions of the plants. Phacelia can successfully reduce *B. cinerea* primary inoculum on vine debris and increase vine debris degradation rate and soil biological activity, compared to a bare-ground control. *B. cinerea* severity can be reduced by up to 10-fold in vine inflorescences at flowering (from 15 to 1.5% inflorescence infected) and by up to 22-fold in grapes at harvest (from 2 to 0.09% bunch area infected) when combining phacelia and mulch. This reduction was attributed partly to reduced primary inoculum but also to probable changes in vine physiology and disease resistance. The same results were seen when using another commonly occurring cover crop, perennial ryegrass (L.) (*Lolium perenne* cv. Kingston, Poaceae) grown in situ as a mulch (Jacometti et al., 2007c). However, it still remains to prove phacellia to be a successful candidate, having all the correct morphological and biological traits required by a plant to be used for these dual purposes within a vineyard.

The deployment of herbivore-induced plant volatiles as a form of habitat modification

The deployment of herbivore-induced plant volatiles (HIPVs) is one type of habitat modification that may increase biological control, an ecosystem service.

HIPVs are a form of induced plant defences that may function both through 'top-down', by attracting natural enemies (Dicke and Bruin, 2001), and 'bottom-up' mechanisms, repelling the herbivore (Dicke et al., 1990). Herbivores feeding (Geervliet et al., 1997) or depositing eggs (Hilker and Meiners, 2002) on the plant induces the plant's HIPV production. Many plant species, such as Lime beans (L.) (*Phaseolus lunatus*, Fabaceae) (Dicke et al., 1990), maize (*Zea mays* L. (Poaceae)) (Turlings et al., 1990) and tomato (L.) (*Lycopersicon esculentum*, Solanaceae) (Thaler, 1999) release HIPV signals, which attract natural enemies of their herbivores. HIPVs can directly attract natural enemies to herbivore-affected plants and/or trigger surrounding plants to start producing their own direct or indirect defences (interplant communication) (Dicke and Bruin, 2001). Other, surrounding undamaged plants can also 'pick-up' on the signal of an impending herbivore attack and activate their defences without actually producing a defence mechanism. These 'primed' plants respond more efficiently once under herbivore attack (Engelberth et al., 2004).

HIPVs can be synthetically produced and deploying these as a habitat manipulation tool within CBC may increase the abundance of natural enemies, consequently increasing biological control of herbivores (Thaler, 1999; Kessler and Baldwin, 2001; James, 2003, 2005; James and Grasswitz, 2005; Simpson et al., 2009; Orre et al., 2010). Extensive work on the effect of deploying multiple HIPVs within grapes, carried out by David James and colleagues at Washington State University, has shown that the deployment of synthetically produced methyl salicylate (MeSA), methyl jasmonate (MeJA) and hexanyl acetate (HA) can increase the abundance of two parasitic wasp genera, possibly by 'signalling' to the plants to produce their own HIPVs (James et al., 2005). Similarly, the abundance of two parasitoids, the encyrtid parasitoids of scale insects (*Metaphycus* sp.) and the egg parasitoids of grape leafhoppers (*Anagrus* sp.), can increase in abundance in response to effects of the three HIPVs (James, 2005). The abundance of five species of predatory insects can be increased in response to MeSA when it is deployed within grape blocks (James and Price, 2004). However, caution is required before the deployment of HIPVs for pest suppression can be considered as a pest management tool as it is possible that effects on the second (herbivore) and fourth (natural enemy of natural enemies) trophic level may occur.

Habitat modification may provide further ecosystem services

Wine makers that produce their wines in a sustainable manner can use habitat modification to promote their business. In New Zealand, this marketing strategy is already used to promote the country as 100% pure New Zealand. The sustainability concept is increasing in importance as a market driver for the country's key export destinations such as the UK and Australia.

In California, with its booming population, the adoption of sustainable wine-growing practices is important for the industry to be able to continue thriving alongside the growing population and at the same time remain able to compete in an increasingly demanding global consumers' market. To encourage and assess the development of sustainable wine growing, California Sustainable Winegrowing

Alliance (CSWA) has, with the help of environmentalists, regulators, university educators and social equity groups, developed a workbook containing 14 chapters of practical guidelines. One of the chapters is encouragingly on ecosystem management. This workbook is one example of how vineyards can improve relations with employees, neighbours and communities and at the same time improve their business, satisfying the growing global consumers' demand for products produced in an environmentally friendly manner. It has been shown that consumers believe that the quality of sustainable wine will be equal to or better than conventionally produced wine and this makes them prepared to pay a higher price for it (Forbes et al., 2009).

More recently, the alternative of adding native plants as a floral resource in the vineyards situation has been developed as a part of the Greening Waipara project in New Zealand. Waipara is one of the larger wine producing regions of the country with over 1800 ha of vineyard development. The project aims to re-establish native New Zealand plants within the Waipara landscape and to increase adoption by growers of sustainable agricultural practices. Using native plants rather than introduced species (such as buckwheat and alyssum) is a novel concept. Using native plants local to the area would not only lead to a potential increase of biological pest control but also an increase in other ecosystem services, such as ecosystem restoration and species conservation. Bringing back some of the fauna that originally inhabited the region may also provide direct services for human enjoyment, such as recreational opportunities and improved landscape aesthetics. New marketing opportunities for the vineyards involved in this project have also arisen through the added value that the 'greening' provides.

So how can native plants increase CBC and how can the planting of native plants in a vineyard setting contribute to ecosystem restoration and species con-servation? Considering that New Zealand has more than 2000 native flowering plant species, there is significant likelihood that some of these plants could be intentionally established (or preserved) to provide resource subsidies to natural enemies. Recent work by Tompkins (2009), within the 'greening Waipara project' in New Zealand, on the deployment of 14 different native plant species beneath Pinot noir vines shows that the endemic plant species *Geranium sessiliflorum* (Simpson and Thomson) (Geraniacea) and *Hebe chathamica* (Cockayne et Allan) (Plantaginaceae) can significantly increase the invertebrate diversity compared to bare earth or ryegrass (Fig. 4.3).

Predation is thought to be one of the most important controlling factors of the light brown apple moth population, accounting for 48–94% of the mortality (HortNet, 2000), with spiders being one of the key predators. When the two plant species were added the abundance of spiders increased significantly compared to bare earth or ryegrass (Table 4.1) (Tompkins, 2010).

These spiders included members of the Theridiidae and Clubionidae families (Tompkins, 2010), which are predators of the light brown apple moth in parts of New Zealand (HortNet, 2000).

Consequently, adding certain species of native groundcover can support a greater diversity and abundance of invertebrate fauna than ryegrass or bare earth. An increase of the 'right' arthropods, such as natural enemies of pest, has the potential to translate into greater biological control. Adding native vegetation

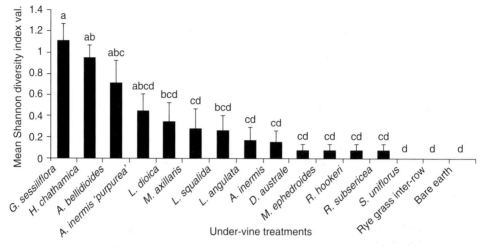

Fig. 4.3 Mean (±SE) Shannon diversity index values of under-vine treatments. Treatments which share letters do not significantly differ using a Tukey-type test $\alpha=0.05$. (From Tompkins, 2010.)

Table 4.1 Density of spiders (Mean (±) SE / 0.04 m²) in different under-vine treatments at three sampling dates (from Tompkins, 2010).

Under-vine treatment	Sampling date[‡]		
	August 2008	January 2009	March 2009
Acaena inermis	0.2±0.13	0.6±0.31	0.7±0.28*
Acaena inermis 'purpurea'	0.6±0.4	0.4±0.24	1.8±0.59*[†]
Anaphalioides bellidioides	0.7±0.3	0.7±0.35	0.9±0.34*
Disphyma australe	0.4±0.22	ns	ns
Geranium sessiliflorum	2.2±0.6 *[†]	1.5±0.61*	3.3±1.03*[†]
Hebe chathamica	1.5±0.27 *[†]	1.8±0.71*	2.8±0.88*[†]
Leptinella dioica	0.3±0.15	0.0002±0.00	0.2±0.13
Leptinella squalida	0.1±0.1	0.4±0.24	0±001
Leptinella angulata	0±0	1.3±0.55*	0.8±0.31*
Muehlenbeckia axillaris	0.8±0.42 *	0.3±0.20	1.2±0.42*[†]
Muehlenbeckia ephedroides	0±0	ns	ns
Raoulia hookeri	0±0	0.3±0.20	0.5±0.22*
Raoulia subsericea	0±0	ns	ns
Scleranthus uniflorus	0.1±0.1	0.1±0.10	0±001
Rye grass inter-row	0.1±0.1	0.4±0.24	0.2±0.13
Bare earth	0±0	0.1±0.10	0±001

*significantly different from bare earth treatment ($P <0.05$).
[†]significantly different from rye grass inter-row treatment ($P <0.05$).
ns = treatment not sampled.
[‡]A general linear mixed model analysis revealed no sampling date by treatment effect ($F_{25,279}=0.61$, $P >0.05$).

to a highly disturbed habitat has the ability to provide both cultural and aesthetical value to the landscape. For example, adding a biodiversity trail to the winery setting may increase the experience in the vineyard, making the visitor feel more connected to the particular winery (Tompkins, 2010). In a country like New Zealand, a biodiversity hotspot, where 80% of the plant species occur only within the country (Wilson, 2004; Meurk et al., 2007) and many of these plants have important associations with the ethics of local iwi, the local Maori people, establishing indigenous species is of both cultural and aesthetic value.

The future

Modifying the vineyard environment to provide both 'indirect ecosystem services', such as increased levels of ecological functions and consequently increased crop productivity, and 'direct ecosystem services' for human enjoyment, such as recreational and aesthetic amenities, has the potential to reverse the association of modern agriculture with biodiversity loss to an association of widespread biodiversity enhancement and increased ecosystem service provision. This chapter touches on some of the more trialled habitat manipulations utilized to enhance ecosystem services as well as on some of the newer, less established ones. Further exploration of the potential for combining multiple habitat manipulation methods may perhaps one day bring back more of the landscape that once surrounded Dionysus in ancient Greece while ensuring vine productivity is retained.

References

Aitken, A.G. and Hawlett, E.W. (2010). *Fresh Facts. New Zealand Horticulture*. New Zealand Institute for Plant and Food Research Limited, Auckland, New Zealand.

Beef + Lamb New Zealand (2010). *Compendium of New Zealand Farm Facts*, 34th edn, April 2010. Beef + Lamb New Zealand Publication No. P10013.

Berndt, L.A. and Wratten, S.D. (2005). Effects of alyssum flowers on the longevity, fecundity and sex ratio of the leafroller parasitoid *Dolichogenidea tasmanica*. *Biological Control*, 32, 65–69.

Berndt, L.A., Wratten, S.D. and Scarratt, S.L. (2006). The influence of floral resource subsidies on parasitism rates of leafrollers (Lepidoptera: Tortricidae) in New Zealand vineyards. *Biological Control*, 37, 50–55.

Carpenter, S.R. and Kitchell, J.F. (1993). *The Trophic Cascade in Lake Ecosystems*. Cambridge University Press, Cambridge.

Dicke, M. and Bruin, J. (2001). Chemical information transfer between plants: back to the future. *Biochemical Systematics and Ecology*, 29, 981–994.

Dicke, M., Sabelis, M.W., Takabayashi, J., Bruin, J. and Posthumus, M.A. (1990). Plant strategies of manipulating predator-prey interactions through allelochemicals: prospects of application in pest control. *Journal of Chemical Ecology*, 16, 3091–3118.

Engelberth, J., Alborn, H.T., Schmelz, E.A. and Tumlinson, J.H. (2004). Airborne signals prime plants against insect herbivore attack. *Proceeding of the National Academy of Science USA*, 101, 1781–1785.

FAO (2009). *Wine Production Tons*. Available at: http://faostat.fao.org/site/636/DesktopDefault.aspx?PageID=636#ancor (accessed November, 2010).

Forbes, S.L., Cohen, D.A., Cullen, R., Wratten, S.D. and Fountain, J. (2009). Consumer attitude regarding environmentally sustainable wine: an exploratory study of New Zealand market place. *Journal of Cleaner Production*, 17, 1195–1199.

Geervliet, J.B.F., Posthumus, M.A., Vet, L.E.M. and Dicke, M. (1997). Comparative analysis of headspace volatiles from different caterpillar-infested or uninfested food plants op *Pieris* species. *Journal of Chemical Ecology*, **23**, 2935–2954.

Girvan, M.S., Bullimore, J., Pretty, J.N., Osborn, A.M. and Ball, A.S. (2003). Soil type is the primary determinant of the composition of the total and active bacterial communities in arable soils. *Applied and Environmental Microbiology*, **69**, 1800–1809.

Gurr, G.M., Scarratt, S., Jacometti, M.A. and Wratten, S. (2007). Management of pests and diseases in New Zealand and Australian vineyards. In: *Biological Control a Global Perspective* (eds C. Vincent, M.S. Goettel and G. Lazarovits), pp. 392–398. CAB International, Oxfordshire.

Gurr, G.M., Wratten, S.D. and Barbosa, P. (2000). Success in conservation biological control of arthropods. In: *Biological Control: Measures of Success* (eds G.M. Gurr and S.D. Wratten), pp. 105–132. Kluwer Academic Publishers, Dordrecht, the Netherlands.

Hilker, M. and Meiners, T. (2002). Induction of plant responses towards oviposition and feeding of herbivorous arthropods: a comparison. *Entomologia Experimentalis et Applicata*, **104**, 181–192.

HortNet (2000). Lightbrown apple moth natural enemies and diseases. Available at: http://www.hortnet.co.nz/key/keys/info/enemies/lba-enem.htm (accessed August 2012).

Jacometti, M., Wratten, S.D. and Walter, M. (2010). Review: Alternatives to synthetic fungicides for Botrytis cinerea management in vineyards. *Australian Journal of Grape and Wine Research*, **16**, 154–172.

Jacometti, M.A., Wratten, S.D. and Walter, M. (2007a). Enhancing ecosystem services in vineyards: using cover crops to decrease botrytis bunch rot severity. *International Journal of Agricultural Sustainability*, **5**, 305–314.

Jacometti, M.A., Wratten, S.D. and Walter, M. (2007b). Management of understorey to reduce the primary inoculum *Botrytis cinerea*: Enhancing ecosystem services in vineyards. *Biological Control*, **40**, 57–64.

Jacometti, M.A., Wratten, S.D. and Walter, M. (2007c). Understorey management increases grape quality, yield and resistance to *Botrytis cinerea*. *Agriculture, Ecosystems and Environment*, **122**, 349–356.

James, D.G. (2003). Synthetic herbivore-induced plant volatiles as field attractants for beneficial insects. *Environmental Entomology*, **32**, 977–982.

James, D.G. (2005). Further field evaluation of synthetic herbivore-induced plan volatiles as attractants for beneficial insects. *Journal of Chemical Ecology*, **31**, 481–495.

James, D.G., Castle, S.C., Grasswitz, T.R. and Reyna, V. (2005). Using synthetic herbivore-induced plant volatiles to enhance conservation biological control: field experiments in hops and grapes. *Second International Symposium on Biological Control of Arthropods*, Davos, Switzerland.

James, D.G. and Grasswitz, T.R. (2005). Synthetic herbivore-induced plant volatiles increase field captures of parasitic wasps. *Biocontrol*, **50**, 871–880.

James, D.G. and Price, T.S. (2004). Field testing of methyl salicylate for recruitment and retention of beneficial insects in grapes and hops. *Journal of Chemical Ecology*, **30**, 1613–1628.

Jonsson, M., Wratten, S., Landis, D., Tompkins, J.-M. and Cullen, R. (2010). Habitat manipulation to mitigate the impacts of invasive arthropod pests. *Biological Invasions*, **12**, 2933–2945.

Kessler, A. and Baldwin, I.T. (2001). Defensive function of herbivore-induced plant volatile emission in nature. *Science*, **305**, 665–668.

Kratz, W. (1998). The bait lamina test: General aspects, applications and perspectives. *Environmental Science and Pollution Research*, **5**, 94–96.

Meurk, C.D., Stancu, C. and Smith, A. (2007). Biodiversity in crisis: a crucial role for business. *Second International Conference on Sustainability Engineering and Science*, Auckland, New Zealand.

Orre, G.U.S., Wratten, S., Jonsson, M. and Hale, R.J. (2010). Effects of an herbivore-induced plant volatile on arthropods from three trophic levels in brassicas. *Biological Control*, **53**, 62–67.

Root, R.B. (1973). Organization of plant-arthopod association in simple and diverse habitats: the fauna of collards (*Brassica oleracea*). *Ecological Monographs*, **43**, 95–120.

Scarratt, S.L. (2005). *Enhancing the Biological Control of Leafrollers (Lepidoptera: Tortricidae) using Floral Resource Subsidies in an Organic Vineyard in Marlborough, New Zealand*. PhD Thesis, Lincoln University, Canterbury, New Zealand.

Schwarzenegger, A., Adams, L.S. and Warmerdam, M.-A. (2010). *Summary of Pesticide use Report Data 2009-Index by Commodity*. California Department of Pesticide Regulation, Sacramento, California, USA.

Simpson, M., Gurr, G., Simmons, A.T., et al. (2009). Synthetic herbivore induced plant volatiles – a tool for enhancing conservation biological control of crop pests. *Third International Symposium on Biological Control of Arthropods*, February 8–13, 2009, p. 624. USDA Forest Service, Christchurch, New Zealand.

Thaler, J.S. (1999). Jasmonate-inducible plant defences cause increased parasitism of herbivores. *Nature*, **399**, 686–688.

Tompkins, J.-M. L. (2009). Endemic New Zealand plants for pest management in vineyards. *Third International Symposium on Biological Control of Arthropods*, pp. 234–245. USDA Forest Service, Christchurch, New Zealand.

Tompkins, J.-M. L. (2010). *Ecosystem Services Provided by Native New Zealand Plants in Vineyards*. Lincoln University, Lincoln, New Zealand.

Turlings, T.C.J., Tumlinson, J.H. and Lewis, W.J. (1990). Exploitation of herbivore-induced plant odors by host-seeking parasitic wasps. *Science*, **250**, 1251–1253.

Tylianakis, J.M., Didham, R.K. and Wratten, S.D. (2004). Improved fitness of aphid parasitoids receiving resource subsidies. *Ecology*, **85**, 658–666.

Wilson, K.-J. (2004). *The Flight of the Huia*. University Press, Christchurch, New Zealand.

Wine Institute (2006). California wine has $51.8 billion economic impact on state and $125.3 billion on the U.S. economy. Available at: http://www.wineinstitute.org/resources/pressroom/120720060 (accessed August 2012).

5

Aquaculture and Ecosystem Services: Reframing the Environmental and Social Debate

Corinne Baulcomb

Scottish Agricultural College, Edinburgh, Scotland

Abstract

With increasing global population and the status of many wild fisheries declining, the development of environmentally and socially sustainable aquaculture is of paramount importance. It is unlikely, however, that the growth of aquaculture can occur without also incurring environmental and social impacts. It is important, however, to assess these impacts in a systematic and holistic way. This chapter proposes that this can be accomplished by: (1) mapping the outputs of aquaculture onto an ecosystem services framework, and (2) pairing said analysis with a life-cycle assessment-based approach to the analysis of the inputs to aquaculture systems.

Introduction

The future of aquaculture is a controversial subject, and one that often elicits highly polarized perspectives (Costa-Pierce and Bridger, 2002; Pillay, 2004; Bostock et al., 2010). Although often not explicitly expressed in economic terms, this controversy ultimately stems from different perceptions of what financial, social and environmental costs and benefits would be implied by various industry development trajectories, as well as from concerns about the true sustainability of these development trajectories. Although it is unlikely that these debates will be resolved at a global level, it is theoretically feasible to do this on more local scales, as long as there is explicit recognition and analysis of the trade-offs that

Ecosystem Services in Agricultural and Urban Landscapes, First Edition. Edited by Steve Wratten, Harpinder Sandhu, Ross Cullen and Robert Costanza.

occur between these three spheres of concern (Williams, 1997; Marintez-Alier, 2001; Bostock et al., 2010).

The concept of ecosystem services can help to facilitate this type of analysis by making the link between humans and the environment explicit, even when this link is not captured in traditional and/or formal markets.[1] Despite this, however, there has been little research that frames the environmental and social debates surrounding aquaculture in the context of ecosystem services. One of the consequences of this is that the potentially negative (and positive) impacts of aquaculture have largely been researched and discussed in isolation from each other, rather than in the context of trade-off analysis, and without sufficient reference to human preferences.

This chapter reframes the debate over the environmental and social impacts of different forms of aquaculture by illustrating how the impacts flowing from the outputs of aquaculture operations can be mapped onto a well-known ecosystem service typology. It is argued that, if done on a site and system-specific basis, this will facilitate the analysis of: (1) the social, environmental and financial impacts of particular aquaculture operations, (2) the trade-offs between them, and (3) the sustainability of those trade-offs. The relevance and complementary nature of life-cycle assessment to this approach is also discussed.

Aquaculture and the environment

Although aquaculture of some form has been practiced for millennia (Stickney, 2009), the debate over its future has become highly polarized. Proponents of aquaculture point to the over-exploitation of marine fisheries, increasing human population, increasing demand for fish (and for protein in general), the need to improve developing world food security and the flexibility of aquaculture design, and argue that aquaculture has an important role to play in the sustainable management of earth's natural resources (Bardach, 1997b; Donaldson, 1997; Shang and Tisdell, 1997; Mathias, 1998; Stickney, 2009). Proponents additionally argue that aquaculture may also have an important role to play in increasing biodiversity, ecosystem restoration and ecosystem protection when implemented correctly (Bardach, 1997a; Pillay, 2004). Complementary to these perspectives is the argument that the negative environmental and social impacts that have resulted from aquaculture to date are not intrinsic to the cultivation of aquatic species, per se, but rather a consequence of poor planning, regulation, education and management. This has the implication that these negative impacts are therefore both correctable and avoidable (Boyd and Schmittou, 1999; Bostock et al., 2010).

Opponents to the continued expansion of aquaculture, however, disagree with this assessment, sometimes vehemently (Boyd and Schmittou, 1999; Costa-Pierce and Bridger, 2002), and environmental concerns have often been raised regarding the following: changes in water quality, sedimentation, threats to wild

---

[1] The definition of, and additional details about, ecosystem services can be found in the earlier chapters of this book.

populations of aquatic organisms, chemical residues, antibiotic use, heavy metal concentrations, the destruction of coastal environments and cropland, noise and air pollution, consumption of resources, threats to mammal and bird populations, threats to biodiversity, the sustainability of feed practices, over-abstraction of water resources, the generation of harmful algal blooms, land and/or coastal access and livelihood conflicts, and the establishment of invasive species (Beveridge et al., 1994; Bardach, 1997a; Corbin and Young, 1997; Shang and Tisdell, 1997; Kautsky et al., 2000; Islam et al., 2004; Stickney, 2009; Liu et al., 2010).

A typology of aquaculture operations and the link to ecosystem services

Discussing the environmental and social impacts of 'aquaculture', however, is mis-leading and largely unhelpful to resolving debates about the future of aquatic species cultivation. 'Aquaculture' is an umbrella term that encompasses a substantial number of permutations of species, environmental, social, economic and cultural contexts, each of which will have its own set of environmental and social impacts that are worth considering explicitly and systematically (Bostock et al., 2010).

Accordingly, aquaculture operations can be categorized in a wide variety of ways. They may be categorized by their location (land-based, fresh water-based, or marine-based). They may be categorized by the intensity of the production (extensive/traditional, semi-intensive, intensive, or super intensive). They may be categorized by the number of species that they feature (monoculture cultivation, polyculture cultivation, or fully integrated multitrophic cultivation). They may be categorized by the species reared, and may be further classified according to whether they feature micro- or macroalgae, aquatic grasses, crustaceans, shell-fish, or finfish (or a mix of thereof). Within each of these general classes of culti-vated organism, there are further divisions based on biology or feed dependence that may be worth considering in certain instances. Finfish, for example, can be described by what they eat (i.e. whether they are omnivorous, carnivorous, her-bivorous), where they feed in the water column, and what temperature range they require, or by whether or not their life cycle has been fully closed. Finally, aquaculture operations can be categorized according to the cultivation system employed (raceways, net pens, cages, ponds, lines, rafts, or recalculating systems) (Bardach, 1997b; Shang and Tisdell, 1997; Yuan, 2007; Stickney, 2009).

When the variety of environment types, social contexts (inclusive of social histories and preferences), and economic systems in which aquaculture may operate is also considered, it goes some way to revealing the complexity of the analysis that is actually required to assess the impacts of any particular aquacul-ture operation. As a result of this complexity, it is likely that current patterns of impact will continue, recognized or not, until a full account is taken of them for particular culturing systems and in particular contexts (Bardach, 1997a, b; Williams, 1997; Pillay, 2004). This is largely why reframing the aquaculture debate in terms of a structured ecosystem services typology is important – it helps to ensure that the range of resulting benefits and costs can be made explicit, organized, assessed and valued.

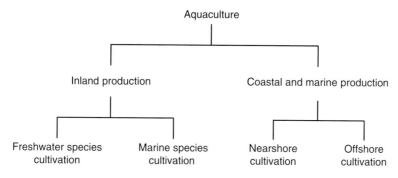

Fig. 5.1 Basic typology differentiating between different types of aquaculture.

For the purposes of illustrating how this may be done, aquaculture operations have, for this chapter, been organized as shown in Fig. 5.1. It should be noted that this is a generic typology that does not distinguish between variables such as cultivation system type, intensity, species type or country location. While all of these factors would be important to consider at the case study level, this typology works well in this context because it allows for a conceptual division to be made between terrestrial and freshwater ecosystem services, on the one hand, and marine ecosystem services on the other.

Because nearly all forms of aquaculture represent, to varying degrees, systems engineered by humans, and because the concept of ecosystem services is derived from the idea that humans benefit in a variety of ways from healthy, functioning ecosystems (Daily, 1997), this chapter puts forth the argument that aquaculture systems should not be equated with natural systems, and that they should instead be considered to be human constructs capable of augmenting or undermining the provision of a suite of ecosystem services by the surrounding environment. One consequence of adopting this perspective is that it requires that assessments of the impact of aquaculture systems on ecosystem services extend well beyond a simple quantification of the amount of food produced by any given aquaculture system. Instead, it requires the consideration of the impact of the following broad attributes of aquaculture production systems on the (local) provision of ecosystem services:

1 the product raised (i.e. the cultured species and the quantity);
2 the culture system infrastructure used (i.e. cages, ponds, etc.);
3 the escapees (i.e. number and species);
4 the fluid emissions produced (i.e. volume and composition/quality of effluent);
5 the gaseous emissions produced (i.e. the volume and composition/quality of gases);
6 the sediment produced (i.e. volume and composition/quality).

These attributes are important because they represent the primary outputs of aquaculture operations, and are therefore those attributes of aquaculture production that interact with the surrounding environment. When combined with a suite of ecosystem services, these attributes create a fairly large 'impact space' that requires assessment. This takes the form of an $m \times n$ matrix where the relevant m aquaculture system attributes are essentially treated as drivers of the

Table 5.1   The TEEB typology of ecosystem services. For details on each service, see de Groot et al., 2010.

Provisioning	Regulating	Habitat	Social–cultural services
1. Food provision	7. Air quality regulation/ purification	16. Maintenance of the life cycles of migratory species by habitat	18. Aesthetic information/ experiences
2. Water provision (quantity)	8. Climate regulation		
3. Raw material provision	9. Moderation of extreme events		19. Opportunities for recreation and tourism
4. Genetic resources provision	10. Regulation of water flows		
5. Medicinal resources provision	11. Waste treatment (including water purification)	17. Maintenance of genetic diversity (including biodiversity) through evolutionary processes	20. Inspiration for culture, art, and design
6. Ornamental resources provision	12. Erosion prevention		21. Spiritual experience
	13. Maintenance of soil fertility		22. Information for cognitive development
	14. Pollination		
	15. Biological Control		

provision of each of the identified n ecosystem service. Assessment of this 'impact space' will facilitate the systematic analysis of the net impacts of a coupled aquaculture–ecological system. If this is additionally combined with marginal economic valuation, assessment of this 'impact space' will facilitate an analysis of the social welfare trade-offs implied by any given aquaculture operation.

In order to illustrate how this might be accomplished, several theoretical case studies are presented in the sections of this chapter that follow below. For the purpose of presenting these case studies, this chapter uses the ecosystem service typology that was assembled for *The Economics of Ecosystems and Biodiversity* (TEEB) project (de Groot et al., 2010), as shown in Table 5.1.[2]

When combined with the aforementioned attributes of aquaculture production, this service typology creates a fairly large 'impact space' that requires assessment (i.e. the 6×22 matrix, as shown in Table 5.2).

The theoretical and illustrative case studies discussed below demonstrate, in a qualitative way, how the matrix shown above might be utilized in four different scenarios. The first pair of case studies demonstrates this through a discussion of

---

[2] This typology evolved out of the ecosystem goods and services typology presented in the 2005 Millennium Ecosystem Assessment, the late twentieth century shift towards ecosystem-based managed, as well as several seminal publications, including de Groot (1992), Costanza et al. (1997) and Daily (1997).

Table 5.2 The aquaculture–ecosystem service 'impact space' derived from the combination of the TEEB ecosystem service typology and the identified aquaculture attributes.

	TEEB ecosystem service typology																					
	Provisioning services							Regulating services									Habitat services		Social–cultural services			
	1	2	3	4	5	6	7	8	9	10	11	12	13	14	15	16	17	18	19	20	21	22
Product																						
Culture system																						
Escapees																						
Effluent																						
Gas emissions																						
Sediment																						

Aquaculture attributes

two contrasting inland production systems, while the second pair does so through a discussion of two contrasting marine-based production systems.

Inland production systems

Overview

Aquaculture operations that are located inland may be placed on parcels of land, in freshwater bodies or in somewhat rarer cases in brackish inland water bodies. While culturing operations that are physically located within freshwater only cultivate freshwater species, aquaculture operations that are located on land may cultivate either freshwater organisms (such as carp) or marine organisms (such as shrimp), depending on the degree of containment maintained between system and the surrounding environment.[3]

Across all the inland production systems, global inland aquaculture operations produced nearly 42 million tons of aquatic organisms in 2010, a number that represents an 21.3 million ton increase from the level of production seen in 2000 (FAO, 2012). The vast majority of this 2010 production (just over 35 million tons) comes from the cultivation of freshwater fish in freshwater environments in Asia. However, there are also substantial quantities of species produced from inland marine and brackish water systems. In 2010, inland aquaculture systems featuring marine and brackish waters produced more than 2.9 million tons of crustaceans, and more than 250000 tons of demersal marine fish (FAO, 2012).

As discussed in the previous section, in order to map the impacts of inland aquaculture production systems onto an ecosystem services typology, the impact space shown in Table 5.2 must be assessed for particular production systems. Once this 'impact space' is filled in for a particular case study, either qualitatively or quantitative, reading both across the rows and down the columns of this table helps to facilitate the analysis of the ecosystem service trade-offs resulting from the pursuit of aquaculture activities. The two hypothetical case studies in this section illustrative this qualitatively for an example of freshwater species production and an example of marine species production.

The first case study illustrates this process for a hypothetical integrated aquaculture–agriculture, carp polyculture system. The focus on carp stems from the fact that more than 23 million tons of carp were cultured around the world in 2010 (FAO, 2012). The focus on this particular type of aquaculture system stems from the argument that these integrated systems show a great deal of promise both in terms of supporting rural livelihoods in developing countries, and in terms of being environmentally sustainable.

The second case study illustrates this process for a hypothetical inland, intensive salt-water-based aquaculture system focused on the production of shrimp. This example was chosen partly because it provides a good contrast with the first case study, and partly because shrimp and prawn species constitute approximately

---

[3]Rao and Kuman (2008), Tal et al. (2009) and Nobre et al. (2010) contain some examples of this type of culture system for species other than shrimp.

94.5% of the more than 2.9 million tons of crustaceans that were cultured inland during 2010 in marine and brackish water (FAO, 2012).

Case study 1: hypothetical integrated agriculture–aquaculture carp polyculture[4]

Integrated agriculture–aquaculture (IAA) systems are compatible with livestock-based production and with crop-based production, and show great potential in ponds, lakes and reservoirs (Mathias, 1998; Pullin, 1998; Nhan et al., 2007; Flores-Nava, 2007; Amilhat et al., 2009; Ahmed et al., 2010). This type of system utilizes both the responsiveness of ponds, including man-made ponds, to farm-derived waste from animals and plants, and the ability of various the carp species to feed throughout the water column (Chen et al., 1998; Duan et al., 1998; Guo et al., 1998; Mathias et al., 1998; Pekar and Olah, 1998; Rahman et al., 1998; Wu et al., 1998; Zhu et al., 1998; Pillay, 2004; Pillay and Kutty, 2005).

For this particular exercise, it is assumed that the system in question consists of an inland, semi-intensive integrated aquaculture–agriculture system with man-made ponds and carp polyculture. It is further assumed that the ponds are sufficiently removed from natural bodies of water to prevent the escape of any farmed fish into wild populations, and that effluent is used for irrigating crops rather than being released directly into proximal sources of freshwater. Given these assumptions and partial culture system specification, a illustration for how the 'impact space' shown in Table 5.2 might be assessed qualitatively on a per aquaculture attribute basis can be found in Table 5.3. Some highlights from this with regard to the four service categories are discussed in more detail below.

Provisioning services

One of the most important provisioning services in terms of this kind of assessment is food, and this service could be affected in a variety of ways by the construction of this type of aquaculture system. When successfully and carefully implemented, the aquaculture system described here would augment the provision of food provided by the agricultural activities on the land both directly and indirectly. The direct part of this augmentation, quite simply, would come from the production of fish. This direct augmentation of the food provision service is particularly beneficial when the culture system is situated on land that is otherwise unsuitable to agricultural or livestock-based activities (Mathias, 1998; Pillay and Kutty, 2005).

The indirect (positive) impact on the food provision service can occur through a variety of pathways. The ability of this type of aquaculture system to use human, animal and crop wastes within the ponds may increase the hygienic conditions on the farm, something that can, in turn, generate more productive farm animals (Uddin et al., 1998). The fish may also, depending on the layout of the system, be able to reduce the insect burden experienced by the local crops, and thereby

---

[4]This same exercise can be repeated for other IAA, intensive, and plant-culture systems as well. For examples of some other systems to which this could be applied, see the following publications: Nhan et al. (2007); Rao and Kumar (2008); Sheng et al. (2009); Tal et al. (2009); Amilhat et al. (2009); Ayer and Tyedmers (2009); Tello et al. (2010); and Ahmed et al. (2010).

Table 5.3 Aquaculture outputs and common potential linkages to ecosystem services. The '+' symbol indicates that the aquaculture output enhances an ecosystem service, where as a '−' indicates that an aquaculture output detracts from an ecosystem service. The symbol '±' indicates that an output of aquaculture may have either positive or negative impacts on an ecosystem service, depending on particular circumstances. This is an illustrative figure for an integrated agriculture–aquaculture, carp polyculture. The pattern of the boxes (and whether they represent positive impacts or negative impacts) will vary according to the particulars of specific culture systems and site-specific variables, and is not exhaustive.

		Provisioning services									Regulating services						Habitat services		Social–cultural services				
		1	2	3	4	5	6	7	8	9	10	11	12	13	14	15	16	17	18	19	20	21	22
Carp polyculture production	Product	+										+				+							+
	Culture system infrastructure		±	−			−		−	+	±						−	−	±	−	+	−	+
	Escapees																						
	Effluent	±										±	+	+		+			−				+
	Gas emissions								−														
	Sediment	+							−			±	+	+		+							+

bolster crop yields (Pillay and Kutty, 2005). The ponds can also be constructed in such a way as to store irrigation water, and the sediment produced in the ponds can be applied as a fertilizer to the land, both of which can further augment the production of food from small parcels of land (Duan et al., 1998 ; Rahman et al., 1998; Ruddle and Prein, 1998; Uddin et al., 1998). These indirect pathways through which an IAA carp polyculture system may augment the local provision of food are especially relevant in poorer contexts where income constraints may prevent or severely limit the use of inorganic fertilizers or pesticides in crop production and animal husbandry.

However, it is important to note that this type of system does not intrinsically lead to the augmentation of the food supply on the farm. Because the system does require land, the culture systems infrastructure has the potential, depending on the availability of land and existing uses of that land, to displace and strain livestock and crop production. This displacement could, depending on the context, ultimately decrease the provision of food from the environment (Satia, 1998). The water requirement of these culture systems may also stress crop and livestock production. This could happen either if local water scarcity is such that keeping the ponds full reduces the amount of water available to crops and livestock, or if the ponds are not properly managed, and if as a result the quality of the effluent is too low for the water to be useable in crop production.

The existence of pathways that can lead to either the augmentation or undermining of the provision of food highlights the importance of both the local context, and the specific details of the culture system to any attempt to assess the 'impact space' in Table 5.4 quantitatively, rather than qualitatively.

Regulating services

This particular type of system also has the ability to impact on the suite of regulations services in a variety of ways. Any emissions from electricity or fuel used in the cultivation of the fish would, for example, detract from the local provision of the climate regulation service. Although this impact is likely to be trivial on a per-farm basis, the effect may become more significant across larger scales of IAA system development. Another service that could feasibly be negatively impacted by this type of system is the 'regulation of water flows' service. This effect would result from a situation where the creation and maintenance of these ponds resulted in decreased water flows to the surrounding environment (Beveridge et al., 1994; Pillay, 2004; Stickney, 2009). In contrast to these potentially negative impacts, however, are the potential positive impacts on regulations services. One such example for this type of aquaculture system relates to the moderation of extreme events. Specifically, this type of culture system may help moderate inland floods through a combination of having potentially lowered the water table (Liu et al., 1998) at the same time that it increased the above ground storage volume through pond construction. Similarly, because both the effluent and the sediment from this system can be used to help facilitate crop and plant growth, they both contribute positively to the erosion prevention service, the soil fertility service and the biological/pest control service.

The interaction of the aquaculture system and the waste treatment/water purification service is more complex, in large part because of the interactions between the effluent and the sediment that is produced by the system. In a man-made IAA system featuring the polyculture of carp, the fish contribute positively to the water quality by feeding throughout the water column. This lowers the overall nutrient burden in the water, and allows the fish to help contain algae and aquatic weed populations (Pillay, 2004; Pillay and Kutty, 2005; Yuan, 2007). However, ponds can become overloaded with pests and nutrients when subject to high levels of excess feed and faecal matter (Pillay, 2004). If effluent is too concentrated, it can become depleted of oxygen and overloaded with bacteria. This stresses and harms the fish (i.e. the main aquaculture products), detracts from the overall systems' ability to process wastes and purify water, and makes that water less useful in terms of its other potential applications (Bardach, 1997a, b; Pillay and Kutty, 2005). Furthermore, if the fish themselves cannot keep the pests in check, a whole variety of chemicals, hormones and antibiotics may be applied to the ponds, all of which has the potential to affect the quality of the effluent and the deposited sediment (Pillay, 2004).

Habitat services

As shown in Table 5.3, there is little interaction between this type of system and the habitat services. In a relatively small, semi-intensive system like the one described above, where the aquaculture production is reasonably isolated from nearby freshwater ecosystems, and where the life cycle of primary species being cultured is fully closed (Pillay and Kutty, 2005), the impact on the habitat services is fairly minimal unless the pond system covers up or replaces an area of land that is particularly crucial to maintaining biodiversity (Beveridge et al., 1994; Pillay, 2004).

Social–cultural services

Similarly, it is likely that the impact on social–cultural services provided by the environment will be small and dominated by the impact of the aquaculture system infrastructure (Pillay, 2004). It is anticipated, for example, that the construction of ponds would have a negative impact on any social–cultural services that involved use of the same land and/or water as would be used by the aquaculture system. However, the very act of creating and running an IAA system would, almost by definition, augment the 'information for cognitive development' in the form of increasing expertise with time.

Case study 2: hypothetical inland marine shrimp cultivation

In order to provide a contrast with the freshwater cultivation system described above, an example of how Table 5.2 may be filled in for a land-based, coastal intensive marine shrimp pond is shown in the Table 5.4.[5] These systems require man-made ponds, a consistent supply of seawater and strict isolation from the surrounding environment. They also require intensive feed and chemical inputs. As with the previous case study, the consequences of this in terms of ecosystem services are explored below.

---

[5] Specific case studies of intensive marine shrimp ponds would likely have a different distribution of ecosystem service impacts. The example presented here is illustrative of how the environmental impacts of marine shrimp ponds can be mapped onto an ecosystem services framework.

Table 5.4 Aquaculture outputs and common potential linkages to ecosystem services. The '+' symbol indicates that the aquaculture output enhances an ecosystem service, where as a '−' indicates that an aquaculture output detracts from an ecosystem service. The symbol '±' indicates that an output of aquaculture may have either positive or negative impacts on an ecosystem service, depending on particular circumstances. This is an illustrative figure for a pond-based intensive marine shrimp system. The pattern of the boxes (and whether they represent positive impacts or negative impacts) will vary according to the particulars of specific culture systems and site-specific variables, and is not exhaustive.

		Provisioning services							Regulating services							Habitat services		Social–cultural services				
	1	2	3	4	5	6	7	8	9	10	11	12	13	14	15	16	17	18	19	20	21	22
Product	±						−								−	−	−					+
Culture system infrastructure		−	−			−		−	+	−	−					−	−	−	−	−	−	+
Escapees																						
Effluent		−	−						−			−	−		−	−		−	−	−	−	+
Gas emissions							−	−														+
Sediment	−						−	−	−		−	−	−		−	−	−	−	−	−	−	+

Shrimp production

Provisioning services

One of the most important differences between this type of system and the previously described system in terms of the impact of the system on the food provision is that the augmentation of local food supplies as a consequence of the production of shrimp is, generally, more questionable than for an integrated IAA system like the one described in Case Study 1. The reasons for this are as follows. Firstly, there is the potential for a conflict between the production of cultured shrimp and the populations of wild, harvestable marine species. The realization of this conflict depends largely on the impact of shrimp fry harvesting on the fry (and subsequently adult) populations non-target species. Secondly, as mentioned previously, there is the potential for pond construction to displace other locally focused food-producing activities. Thirdly, because shrimp cultivation is often focused primarily on export markets, there is the potential for shrimp cultivation to generate a trade-off between local food security and company profits (Corbin and Young, 1997; Williams, 1997; Islam et al., 2004; Gunawardena and Rowan, 2005). This is a critically important difference between the two systems highlighted in these case studies in terms of how they relate to the provision of local ecosystem services.

Regulating, habitat and social–cultural services

Looking at the rest of the table, it is clear that many of the potential areas of ecosystem service augmentation identified in the IAA system do not exist in the intensive land-based marine shrimp aquaculture. The facts that shrimp ponds must be built on land (something that pre-empts other social or environmental uses of the land), and that they depend on saltwater (which prevents the re-use of the water or sediment in the context of agriculture and makes disposal of the effluent more difficult) are the two factors largely responsible for the differences shown between Tables 5.3 and 5.4. Additional factors impacting on how inland intensive shrimp monoculture maps onto ecosystem services include the high temperature requirements of shrimp, the dependence of these systems on manufactured feed, and the fact that the conversion of land to shrimp farming often involves the semipermanent conversion of mangroves to less diverse and productive habitat types (Beveridge et al., 1994; Bardach, 1997a; Gupta, 1998; Liu et al., 1998; Pillay, 2004; Gunawardena and Rowan, 2005; Liu et al., 2010).

Taken together, and in particular contrast to the system shown in Table 5.3, this highlights that this particular type kind of aquaculture exists much more in contrast to the surrounding environment. All of these factors listed above imply the potential for significant trade-offs between the provision of a single part of a single ecosystem service (i.e. shrimp), and a wide array of other ecosystem services (and therefore benefits to humans). Although this general idea of trade-offs has been widely vocalized by the critics of intensive shrimp farming, framing these trade offs in terms of changes to the provision of specific ecosystem services, which can then be connected to specific human benefits, helps to make the link between these trade-offs and human well-being explicit.

Marine and coastal-based production systems

Overview

In the context of marine environments, aquaculture operations can be placed in nearshore environments or offshore environments. Currently, nearshore environments host significantly more aquaculture production than do offshore areas, though, increasingly, cost-effective ways are being found to expand the cultivation of marine species in offshore areas (Muir, 2000; Eklof et al., 2006; Corbin, 2007). The reasons for this are several. Firstly, human use of nearshore marine environments is much more developed and intensive than our corresponding use of offshore marine environments. Shipping lanes, recreation zones, tourism and conservation areas, for example, are already well established in many places, and this can make it difficult to situate aquaculture operations in these environments. Secondly, aquaculture operations are themselves very sensitive to incoming levels of pollution to which coastal areas are subject.

This is particularly relevant when culture operations are intensive and cultured organisms are grown in dense populations (Stickney, 1997; Pillay, 2004), as this exposes already stressed organisms to entire catchments worth of nutrient/chemical-rich run-off (Bardach, 1997a). The prospect of offshore aquaculture is therefore appealing because it largely avoids these issues, but has had limited implementation thus far because it poses substantial technical and economic challenges (Stickney, 1997; Muir, 2000).

Across all the nearshore and offshore cultivation systems, the global marine and coastal-based aquaculture efforts produced 37.1 million tons of aquatic organisms in 2010 (FAO, 2012). This is reasonably close to the global freshwater production statistics presented earlier, and represents a 15.8 million ton increase from 2000 production levels (FAO, 2012). As is the case with inland aquaculture production, Asia is the dominant producer in marine and coastal environments, generating nearly 90% of the world's production in 2010. In contrast to inland aquaculture efforts, however, a significant portion of the coastal and nearshore aquaculture focuses on the culturing of aquatic plants. The world cultured nearly 19 million tons of aquatic plants in 2010, a quantity that represents approximately 51% of the total 2010 coastal and marine aquaculture tonnage. Other types of aquatic organisms that are cultured in significant quantities include molluscs, diadromous fishes and marine fishes (FAO, 2012).

As discussed in the previous section, the impact space shown in Table 5.2 must be assessed for particular production systems. The two hypothetical case studies in this section illustrate this qualitatively for an example of nearshore species production and an example of offshore marine species production.

The first case study illustrates the process of re-framing the potential environmental impacts of aquaculture in terms of ecosystem service impacts for a hypothetical intensive, raft-based bivalve culture system. This case study was selected in part because nearly 14 million tons of molluscs were cultured around the world in 2010 (FAO, 2012), and in part because it is not uncommon for shellfish farming to be treated as if it was, inherently and without respect to intensity, environmentally positive (or at least neutral) (SARF, 2008). This case study will highlight that such an assumption may obscure the wider and complex range of environmental impacts that can stem from intensive shellfish farming.

The second case study illustrates this process for a hypothetical 'best-case' scenario for the development of offshore aquaculture. This topic was chosen because there has been increasing interest in the development of this type of aquaculture (James and Slaski, 2006).

Case study 3: hypothetic nearshore, intensive and raft-based shellfish cultivation

In order to illustrate how the outputs of nearshore aquaculture may be mapped onto an ecosystem services framework, this case study focuses on a hypothetical, intensive, raft-based shellfish culture system. It is assumed for the purposes of this illustration that the species in question is native to the location where the aquaculture system is situated, and that this type of cultivation is not new to the area in question. The consequences of this type of system in terms of the provision of a suite of ecosystem services are shown in Table 5.5, and discussed further below.

Provisioning services
As with the other systems discussed, the provision of food is one of the primary objectives of this type of system. The extent to which the food provisioning service of the nearshore ecosystem is augmented by the cultivation of shellfish, however, depends on the ecosystem dynamics of the environment in which the raft is situated. It is true that this service may be augmented substantially purely from the existence of the culturing system. One raft can culture hundreds of thousands of individual shellfish, and nearshore environments may house multiple rafts (Pillay, 2004). However, because shellfish are filter-feeding organisms, they may impact on the food provisioning service in other, less straightforward ways, and their mere presence does not necessarily guarantee an augmentation of the provision of food. Take for example, a situation in which pollution from other land and marine-based activities generates harmful coastal algal blooms that contaminate the shellfish rafts. Although those particular mussels may not suffer ill effects from the algal bloom, they may not end up being suitable for human consumption and, as a consequence, the aquaculture system would fail to augment the provision of food.

The sedimentation that occurs as a consequence of these rafts, however, may undermine a variety of ecosystem services, including the provision of food. The reason for this is the biological reality that shellfish only retain 35–40% of what they filter from the water (Pillay, 2004). As Pillay (2004) points out, this means that a full raft of more than 400 000 oysters will concentrate and deposit more than 21 tons of sediment on the benthos beneath and 'down stream' from the raft over the course of a year. Depending on the local current structure, this sedimentation may happen at a much faster rate than the environment can remove it. In turn, this can lead to a wide host of localized and negative environmental consequences that have the potential to undermine the provision of food (and other ecosystem services) from nearshore environments. These negative

Table 5.5 Aquaculture outputs and common potential linkages to ecosystem services. A '+' indicates that the aquaculture output enhances or augments an ecosystem service, where as a '−' indicates that an aquaculture output detracts from an ecosystem service. The symbol '±' indicates that an output of aquaculture may have either positive or negative impacts on an ecosystem service, depending on particular circumstances. This is an illustrative figure for an intensive, raft-based bivalve culture system. The pattern of the boxes (and whether they represent positive impacts or negative impacts) will vary according to the particulars of specific culture systems and site-specific variables, and is not exhaustive.

		Provisioning services							Regulating services								Habitat services		Social–cultural services				
		1	2	3	4	5	6	7	8	9	10	11	12	13	14	15	16	17	18	19	20	21	22
Nearshore saltwater-based cultivation	Product	+		+			+					+											+
	Culture System Infrastructure							+		+	±								±	±	±	±	+
	Escapees																						
	Effluent	+										+					+	+	+	+	+	+	+
	Gas Emissions							−															
	Sediment	−						−				−	+			−	−	−	−	−	−	−	+

consequences include the abandonment of the seafloor by wild shellfish, increased production of hydrogen sulphide and methane gases, the production of aqueous ammonia, a substantial decrease in the water column's dissolved oxygen content, altered food chain dynamics, decreased pest control abilities and fish kills (Pillay, 2004; Ferreira et al., 2009). This demonstrates that decisions regarding the level of desired aquaculture production should be made in reference to estimations of both the local environmental carrying capacity and current structures, and must treat the culturing of shellfish as an activity that both affects and is affected by complex ecosystem characteristics (Ferreira et al., 2009; Vaudrey et al., 2009).

Regulating services

The aforementioned summary of the potential negative consequences of the intensive mussel raft-induced sedimentation listed the types of changes to a variety of environmental variables that could decrease the food provision of the local system. Although these variables do impact on the provision of food, several of them may also be considered to be regulating services, such waste treatment and biological control, in their own right, many of which are undermined by the sedimentation that results from the presence of the intensive mussel raft. The impact of this type of culture system is not necessarily entirely negative in terms of its impact on regulating services. The ability of mussels to filter pollutants from the water column, for example, augments the waste-processing service of the environment and it is possible that, under certain circumstances, the sedimentation could counteract some erosion. The fact that there are several potentially contradictory responses of regulating services to the presence of an intensive raft-based mussel culture system underlines the importance of carefully exploring the trade-offs implied by any given culture system in any given ecological context.

Habitat services

In terms of habitat services, the impact of the mussels on the habitat services takes the form of a trade-off between the largely positive consequences stemming from improved water quality, and potentially very negative consequences of increased sedimentation and smothering. The pre-existing biophysical characteristics of the location over which the shellfish rafts are situated and the intensity of the cultivation will largely determine whether the potential positive or the potential negative contributions of the intensive shellfish aquaculture culture to the provision of the habitat services are realized.

Social–cultural services

The impact of mussels on the cultural services also takes the form a trade-off between the largely positive consequences of cleaner coastal waters and the negative impacts of sedimentation. Boaters, recreational fishermen, swimmers and other users of coastal waters will benefit directly from an improvement in water quality that results from the instalment of significant numbers of filter feeding organisms. If the correct balance is not achieved, however, and the sedimentation from the shellfish rafts overwhelms the provision of services such as the waste treatment service or the food provision service, then cultural and recreational uses of coastal waters may be impaired, rather than improved as a consequence of shellfish aquaculture. Additionally, it is important to note that while the impact of the culture system's infrastructure on the provision of social–cultural services may be fairly neutral from the perspective of shore-based activities, it may be limit certain water-based recreational activities, and may also affect the aesthetics of the nearshore zone (Corbin and Young, 1997; Stickney, 1997).

Case study 4: hypothetical 'best-case' offshore aquaculture cultivation

As there is insufficient information on open ocean/offshore ecosystem services and offshore aquaculture to make links through an illustrative case study, the bottom half of Table 5.2 is filled to represent an optimistic, best-case scenario for offshore aquaculture. If such a system were to come into existence, it would involve the offshore installation of floating, semimobile sea cages, and would allow for the large scale culturing of native marine species. Such a system might, in a best-case scenario, map onto ecosystem services typology as shown in Table 5.6 below.

Provisioning, regulating, habitat and social–cultural services
In theory, offshore aquaculture can increase the provisioning of sea food, be integrated into some kind of polyculture system, increase local concentrations of marine flora and fauna, and be used to augment the population of native species whose numbers are in decline (Stickney, 1997; Corbin, 2007; Soto and Jara, 2007; Taylor, 2009; Troell et al., 2009). Furthermore, offshore aquaculture facilities will likely cause less damaging effluent and sediment-related damages on the surrounding environment (James and Slaski, 2006), and if out of site of the shore, the impacts on the social–cultural services will be negligible. Achieving this particular picture of offshore aquaculture, however, is contingent upon a wide variety of assumptions that, as James and Slaski (2006) points out, have yet to be definitively tested.

The value of a complementary life-cycle approach

Performing this kind of analysis and framing the outputs of aquaculture in terms of their augmentation of, or detraction from, the provision of various ecosystem services has several advantages over simply discussing the potential environmental impacts of various aquaculture operations. Once the relevant impacts are identified and categorized as shown in Tables 5.4–5.6, they may be quantified. This quantification would allow for an assessment of the net impact of the aquaculture operation on the local environment (in the form of ecosystem services), and will help make clear the trade-offs involved with utilizing a particular production system in a particular context. Crucially, this also facilitates economic assessment. There are well-established (although not uncontested) methodologies for valuing changes in the flow of benefits that result from changes in the delivery of ecosystem services. These methods can use both market-based and

Table 5.6 Aquaculture outputs and common potential linkages to ecosystem services. A '+' indicates that the aquaculture output enhances or augments an ecosystem service, where as a '−' indicates that an aquaculture output detracts from an ecosystem service. The symbol '±' indicates that an output of aquaculture may have either positive or negative impacts on an ecosystem service, depending on particular circumstances. This is an illustrative figure for a possible best case offshore finfish-polyculture system. The pattern of the boxes (and whether they represent positive impacts or negative impacts) will vary according to the particulars of specific culture systems and site-specific variables, and is not exhaustive.

		Provisioning services									Regulating services						Habitat services		Social–cultural services				
		1	2	3	4	5	6	7	8	9	10	11	12	13	14	15	16	17	18	19	20	21	22
Offshore saltwater-based cultivation	Product	+	+	+			+					+				+	+	+					+
	Culture system infrastructure	+		+			+										+	±					+
	Escapees	+					+					+				+		+					+
	Effluent																						+
	Gas emissions									−													+
	Sediment																						+

non-market-based information and can therefore be used to help capture the social welfare impacts of aquaculture operations in addition to the aforementioned environmental impacts.

There are limits to this approach, however. Although aquaculture is an industry that closely couples its production and outputs with onsite or proximal ecological processes, the *inputs* to aquaculture also may have non-trivial, negative environmental or social consequences that are important to consider. In making decisions about the inputs to aquaculture, trade-offs between these different environmental, social and industrial features are being implicitly made, but often not explicitly and systematically considered. In many instances, this is a consequence either of these environmental costs being incurred in another part of the world than an aquaculture operation, or a consequence of these impacts affecting the environment and society on a different scale than the outputs of an aquaculture operation (Karakassis, 1998; Ayer and Tyedmers, 2009). This type of impact, therefore, cannot be analysed by the output-focused framework presented here.

One means of addressing this limitation is to couple the framework presented here for analysing the outputs of aquaculture with a life-cycle assessment (LCA) approach to the main material inputs/throughputs of particular aquaculture operations (i.e. physical materials, energy, water, fry/fingerlings and feed) (Pelletier and Tyedmers, 2008; Aubin and Van der Werf, 2009).

Research that has applied an LCA approach to aquaculture inputs, and aquaculture feed in particular, has yielded some surprising results that demonstrate the utility of this type of analysis in the context of understanding the relationship that various aquaculture inputs have with the environment. For example, Pelletier and Tyedmers (2007) used LCA to assess the relative environmental impact of three alternative varieties of organic salmon feed compared to the impact of a conventional feed. Their research showed that the cradle-to-mill gate impact of salmon feed with organic crop-derived and fisheries by-product-based ingredients had a larger environmental impact across six categories of impact (energy use, global warming potential, marine aquatic ecotoxicity potential, acidification potential, eutrophication potential and biotic resource use), than did the conventional feed (Pelletier and Tyedmers, 2007). In another telling study, Pelletier and Tyedmers demonstrated that, counter to contemporary intuition on the benefits of using plant-derived inputs to aquaculture, using certain crop-derived inputs to tilapia feed had an environmental impact across five impact categories equivalent to that of the fish-derived inputs to that same feed (Pelletier and Tyedmers, 2010). These kinds of result are important because they allow for the impacts of well-intentioned, but insufficiently scrutinized choices to be analysed, particularly if these impacts occur in different environments, or on a different spatial–temporal scale than a particular aquaculture operation.

Conclusion

With increasing global population, the status of an increasing number of wild fisheries is threatened or declining, the development of environmentally and socially sustainable aquaculture is of paramount importance. Although the potential for aquaculture operations to have negative impacts on the environment is real, discussing these impacts out with a framework that allows for the systematic analysis of both the ecological and social trade-offs between different, overlapping impacts not only fails to provide useful information to decision-making processes, but also can be misleading and counter productive to improving the sustainability these operations.

Mapping the outputs of aquaculture production on a site-by-site, case-by-case basis onto a well-structured ecosystem services typology facilitates both a rigorous analysis of the impacts of the outputs from aquaculture production systems, and an economic valuation of those impacts by various stakeholder groups. Information on the impacts in this format can feed directly into decision-making processes and can help communities, scientists and governments decide how best to pursue the cultivation of aquatic organisms.

On its own, however, this type of analysis does not take into account the environmental or social impacts of the inputs that are required to support aquaculture production. As these inputs may incur social or environmental costs elsewhere on the planet or on a different scale than would otherwise be the focus of aquaculture operations, an LCA approach can be utilized to take account of and to characterize these impacts. This will complement the analysis of aquaculture in the context of ecosystem services. When combined, both types of analysis will provide a more complete and useful picture of the impacts of aquaculture than would less well-structured debates about generic potential environmental or social impacts of aquaculture expansion.

Ultimately, what is required is that it be explicitly recognized that aquaculture has both industrial and ecological components, and that as a consequence of this, complementary, structured environmental and social analysis be applied accordingly in order to facilitate both a truer understanding of the costs and benefits of aquaculture production, and the adjustment of culturing techniques to increase their sustainability.

References

Ahmed, N., Allison, E.H., and Muir, J.F. (2010). Rice fields to prawn farms: a blue revolution in southwest Bangladesh? *Aquaculture International*, 18, 555–574.

Amilhat, E., Lorenzen, K., Morales, E.J., Yakupitiyage, A., and Little, D.C. (2009). Fisheries production in southeast Asian Farmer Managed Aquatic Systems (FMAS). II: diversity of aquatic resoruces and management impacts on catch rates. *Aquaculture*, 298, 57–63.

Aubin, J. and van der Werf, H.M.G. (2009). Fish farming and the environment: a life cycle assessment approach [English abstract]. *Cahiers Agricultures*, 18, 220–226.

Ayer, N.W. and Tyedmers, P.H. (2009). Assessing alternative aquaculture technologies: life cycle assessment of salmonid culture systems in Canada. *Journal of Cleaner Production*, 17, 362–373.

Bardach, J.E. (1997a). Aquaculture, pollution, and biodiversity. In: *Sustainable Aquaculture* (ed. J.E. Bardach), pp. 87–99. John Wiley and Sons, Inc., New York.

Bardach, J.E. (1997b). Fish as food and the case for aquaculture. In: *Sustainable Aquaculture* (ed. J.E. Bardach), pp. 1–14. John Wiley and Sons, Inc., New York.

Beveridge, M.C.M., Ross, L.G. and Kelly, L.A. (1994). Aquaculture and biodiversity. *Ambio*, 23, 497–502.

Bostock, J., McAndrew, B., Richards, R., et al. (2010). Aquaculture: global status and trends. *Philosophical Transactions of the Royal Society B*, **365**, 2897–2912.

Boyd, C.E. and Schmittou, H.R. (1999). Achievement of sustainable aquaculture through environmental management. *Aquaculture Economics and Management*, **3**, 59–69.

Chen, H-L., Charles, A.T. and Hu, B-T. (1998). Chinese integrated fish farming. In: *Integrated Fish Farming: Proceedings of a Workshop on Integrated Fish Farming Held in Wuxi, Jiangsu Province, People's Republic of China October 11–15, 1994* (eds J.A. Mathias, A.T. Charles and H. Baotong), pp. 97–109. CRC Press LLC, Boca Rotan.

Corbin, J.S (2007). Marine aquaculture: today's necessity for tomorrow's seafood. *Marine Technology Society Journal*, **41**, 16–23.

Corbin, J.S. and Young, L.G.L. (1997). Planning, regulation, and administration of sustainable aquaculture. In: *Sustainable Aquaculture* (ed. J.E. Bardach), pp. 201–233. John Wiley and Sons, Inc., New York.

Costa-Pierce, B.S. and Bridger, C.J. (2002). The role of marine aquaculture facilities as habitats and ecosystems. In: *Responsible Marine Aquaculture* (eds R.R. Stickney and J.P. McVey), pp. 105–144. CABI Publishing, Wallingford.

Costanza, R., d'Arge, R., de Groot, R., et al. (1997). The value of the world's ecosystem services and natural capital. *Nature*, **387** (6630), 253–260.

Daily, G. (1997). *Nature's Services: Societal Dependence on Natural Ecosystems*. Island Press, Washington, DC.

de Groot, R. (1992). *Functions of Nature: Evaluation of Nature in Environmental Planning, Management and Decision-Making*. Wolders Noordhoff BV, Groningen.

de Groot, R., Fisher, B., Christie, M., et al. (2010). Integrating the ecological and economic dimensions in biodiversity and ecosystem service valuation. In: *The Economics of Ecosystems and Biodiversity: Ecological and Economic Foundations Draft Version*. Availabl eat: http://www.teebweb.org/EcologicalandEconomicFoundationDraftChapters/tabid/29426/Default.aspx (accessed February 2011).

Donaldson, E.M. (1997). The role of biotechnology in sustainable aquaculture. In: *Sustainable Aquaculture* (ed. J.E. Bardach), pp. 101–126. John Wiley and Sons, Inc., New York.

Duan, D.-X., Liu, S.-Y., Su, J.-S., Zhou, G.-S., and Zhang, G.-P. (1998). Fish farming integrated with livestock and poultry in China. In: *Integrated Fish Farming: Proceedings of a Workshop on Integrated Fish Farming Held in Wuxi, Jiangsu Province, People's Republic of China October 11–15, 1994* (eds J.A. Mathias, A.T. Charles and H. Baotong), pp. 73–82. CRC Press LLC, Boca Rotan.

Eklof, J.S., de la Torre-Castro, M., Nilsson, C. and Ronnback, P. (2006). How do seaweed farms influence local fishery catches in a seagrass-dominated setting in Chwaka Bay, Zanzibar? *Aquatic Living Resources*, **19**, 137–147.

FAO (2012). *Global Aquaculture Production 1950–2010*: Online Query. Available 'A-1: World Aquaculture Production by Inland and Marine Waters.' Available at: http://www.fao.org/fishery/statistics/global-aquaculture-production/query/en. (accessed March 2012).

Ferreira, J.G., Sequeira, A., Hawkins, A.J.S., et al. (2009). Analysis of coastal and offshore aquaculture: application of the farm model to multiple systems and shellfish species. *Aquaculture*, **289**, 32–41.

Flores-Nava, A. (2007). Feeds and fertilizers for sustainable aquaculture development: a regional review for Latin America. In: *Study and Analysis of Feeds and Fertilizers for Sustainable Aquaculture Development* (eds M.R. Hasan, T. Hecht, S.S. de Silva, and A.G.J. Tacon). FAO, Rome.

Gunawardena, M. and Rowan, J.S. (2005). Economic valuation of a mangrove ecosystem threatened by shrimp aquaculture in Sri Lanka. *Environmental Management*, **36**, 535–550.

Guo, X.-Z., Fang, X.-Z., Xie, J., Yu, T.-B. and Zhang, W.-Y. (1998). Bacterial productivity in intensely cultured high-yield fish pond using [3h]Tdr tracing analysis techniques. In: *Integrated Fish Farming: Proceedings of a Workshop on Integrated Fish Farming Held in Wuxi, Jiangsu Province, People's Republic of China October 11–15, 1994* (eds J.A. Mathias, A.T. Charles and H. Baotong), pp. 137–145. CRC Press LLC, Boca Rotan.

Gupta, M.V. (1998). Social and policy issues involved in adoption of integrated agriculture-aquaculture-livestock production systems in Bangladesh. In: *Integrated Fish Farming: Proceedings of a Workshop on Integrated Fish Farming Held in Wuxi, Jiangsu Province, People's Republic of China October 11–15, 1994* (eds J.A. Mathias, A.T. Charles and H. Baotong), pp. 229–244. CRC Press LLC, Boca Rotan.

Islam, M.S., Wahab, M.A. and Tanaka, M. (2004). Seed supply for coastal brackish water shrimp farming: environmental impacts and sustainability. *Marine Pollution Bulletin*, **48**, 7–11.

James, M.A. and Slaski, R. (2006). *Appraisal of the Opportunity for Offshore Aquaculture in UK Waters*. Report of Project FC0934, commissioned by Defra and Seafish from FRM Ltd. Available at: http://www.seafish.org/media/Publications/Offshore_Aquaculture_Compiled_Final_Report.pdf (accessed March 2012).

Karakassis, I. (1998). Aquaculture and coastal marine biodiversity. *Oceanis. Serie de documents Oceanographiques*, **24**, 271–286.

Kautsky, N., Ronnback, P., Tedengren, M. and Troell, M. (2000). Ecosystem perspectives on management of disease in shrimp pond farming. *Aquaculture*, **191**, 145–161.

Liu, S.-Y., Yang, L.-B., Duan, D.-X., Gu, H.-L. and Zhang, J.-L. (1998). Using integrated fish farming to reclaim low-lying, saline-alkali land along the Yellow River in China. In: *Integrated Fish Farming: Proceedings of a Workshop on Integrated Fish Farming Held in Wuxi, Jiangsu Province, People's Republic of China October 11–15, 1994* (eds J.A. Mathias, A.T. Charles and H. Baotong), pp. 359–367. CRC Press LLC, Boca Rotan.

Liu, Y.-Y., Wang, W.-N., Ou, C.-X., Yuan, J.-X., Wang, A.-L., Jiang, H.-S. and Sun, R. (2010). Valuation of shrimp ecosystem services: a case study in Leizhou City, China. *International Journal of Sustainable Development and World Ecology*, **17**, 217–224.

Marintez-Alier, J. (2001). Ecological conflicts and valuation: mangroves versus shrimps in the late 1990s. *Environment and Planning C-Government and Policy*, **19**, 713–728.

Mathias, J. (1998). The importance of integrated fish farming to world food supply. In: *Integrated Fish Farming: Proceedings of a Workshop on Integrated Fish Farming Held in Wuxi, Jiangsu Province, People's Republic of China October 11–15, 1994* (eds J.A. Mathias, A.T. Charles and H. Baotong), pp. 3–18. CRC Press LLC, Boca Rotan.

Mathias, J.A., Charles, A.T. and Baotong, H. (1998). *Integrated Fish Farming: Proceedings of a Workshop on Integrated Fish Farming Held in Wuxi, Jiangsu Province, People's Republic of China October 11–15, 1994*. CRC Press LLC, Boca Rotan.

Muir, J.F. (2000). The potential for offshore mariculture. In: *Mediterranean Offshore Mariculture* (eds J. Muir and B. Basurco), pp. 19–24. CIHEAM, Zaragoza, Spain.

Nhan, D.K., Phong, L.T., Verdegem, M.J.C., Duong, L.T., Bosma, R.H., and Little, D.C. (2007). Integrated freshwater aquaculture, crop and livestock production in the Mekong Delta, Vietnam: determinants and the role of the pond. *Agricultural Systems*, **94**, 445–458.

Nobre, A.M., Robertson-Andersson, D., Neori, A. and Sankar, K. (2010). Ecological–economic assessment of aquaculture options: comparison between abalone monoculture and integrated multi-trophic aquaculture of abalone and seaweeds. *Aquaculture*, **306**, 116–126.

Pekar, F. and Olah, J. (1998). Fish pond manuring studies in Hungary. In: *Integrated Fish Farming: Proceedings of a Workshop on Integrated Fish Farming Held in Wuxi, Jiangsu Province, People's Republic of China October 11–15, 1994* (eds J.A. Mathias, A.T. Charles and H. Baotong), pp. 163–177. CRC Press LLC, Boca Rotan.

Pelletier, N. and Tyedmers, P. (2007). Feeding farmed salmon: is organic better? *Aquaculture*, **272**, 399–416.

Pelletier, N. and Tyedmers, P. (2008). Life cycle considerations for improving sustainability assessments in seafood awareness campaigns. *Environmental Management*, **42**, 918–931.

Pelletier, N. and Tyedmers, P. (2010). Life cycle assessment of frozen tilapia fillets from Indonesian lake-based and pond-based intensive aquaculture systems. *Journal of Industrial Ecology*, **14**, 467–481.

Pillay, T.V.R. (2004). *Aquaculture and the Environment*. Blackwell Publishing, Ltd., Oxford.

Pillay, T.V.R. and Kutty, M.N. (2005). *Aquaculture Principles and Practices*. Blackwell Publishing Ltd., Oxford.

Pullin, R.S.V. (1998). Aquaculture, integrated resources management and the environment. In: *Integrated Fish Farming: Proceedings of a Workshop on Integrated Fish Farming Held in Wuxi, Jiangsu Province, People's Republic of China October 11–15, 1994* (eds J.A. Mathias, A.T. Charles and H. Baotong), pp. 19–43. CRC Press LLC, Boca Rotan.

Rahman, M.A., Uddin, M.S., Hossain, M.A., Das, G.B., Mazid, M.A. and Gupta, M.V. (1998). Polyculture of carps in integrated broiler-cum-fish farming systems. In: *Integrated Fish Farming: Proceedings of a Workshop on Integrated Fish Farming Held in Wuxi, Jiangsu Province, People's Republic of China October 11–15, 1994* (eds J.A. Mathias, A.T. Charles and H. Baotong), pp. 83–93. CRC Press LLC, Boca Rotan.

Rao, G.S. and Kumar, R.N. (2008). Economic analysis of land-based production of cultured marine pearls in India. *Aquaculture Economics and Management*, **12**, 130–144.

Ruddle, K. and Prein, M. (1998). Assessing potential nutritional and household economic benefits of developing integrated farming systems. In: *Integrated Fish Farming: Proceedings of a Workshop on Integrated Fish Farming Held in Wuxi, Jiangsu Province, People's Republic of China October 11–15, 1994* (eds J.A. Mathias, A.T. Charles and H. Baotong), pp. 111–121. CRC Press LLC, Boca Rotan.

SARF (2008). *Scoping Study of Appropriate EIA Trigger Thresholds for Shellfish Farms and other Non-fish Farm Aquaculture*. Scottish Aquaculture Research Forum. Final Report 9S2511. Available at: http://www.sarf.org.uk/cms-assets/documents/28812-211733. sarf-031-final-report-13sept08.pdf (accessed March 2012).

Satia, B.P. (1998). The integration of an aquaculture-poultry production system in Cameroon. In: *Integrated Fish Farming: Proceedings of a Workshop on Integrated Fish Farming Held in Wuxi, Jiangsu Province, People's Republic of China October 11–15, 1994* (eds J.A. Mathias, A.T. Charles and H. Baotong), pp. 245–255. CRC Press LLC, Boca Rotan.

Shang, Y.C. and Tisdell, C.A. (1997). Economic decision making in sustainable aquacultural development. In: *Sustainable Aquaculture* (ed. J.E. Bardach), pp. 127–148. John Wiley and Sons, Inc., New York.

Sheng, L., Zong-Xiang, G., Huai-Yu, Y. and Zheng-Yong, T. (2009). Evaluation of cultural service value of aquaculture pond ecosystem: a case study in water conservation area of Shanghai [English abstract]. *Yingyong Shengtai Xuebao*, **20**, 3075–3083.

Soto, D. and Jara, F. (2007). Using natural ecosystem services to diminish salmon-farming footprints in southern Chile. In: *Ecological and Genetic Implications of Aquaculture Activities* (Reviews: Methods and Technologies in Fish Biology and Fisheries) (ed. T.M. Bert), pp. 459–475. Springer, Dordrecht.

Stickney, R.R. (1997). Offshore mariculture. In: *Sustainable Aquaculture* (ed. J.E. Bardach), pp. 53–86. John Wiley and Sons, Inc., New York.

Stickney, R.R. (2009). *Aquaculture: An Introductory Text*. Cambridge University Press, Cambridge.

Tal, Y., Schreier, H.J., Sowers, K.R., Stubblefield, J.D., Place, A.R. and Zohar, Y. (2009). Environmentally sustainable land-based marine aquaculture. *Aquaculture*, **286**, 28–35.

Taylor, D.A. (2009). Aquaculture navigates through troubled waters. *Environmental Health Perspectives*, **117**, A252–A255.

Tello, A., Corner, R.A. and Telfer, T.C. (2010). How do land-based salmonid farms affect stream ecology? *Environmental Pollution*, **158**, 1147–1158.

Troell, M., Joyce, A., Chopin, T., Neori, A., Buschmann, A.H. and Fang, J-G. (2009). Ecological engineering in aquaculture: potential for integrated multi-trophic aquaculture (IMTA) in marine offshore systems. *Aquaculture*, **297**, 1–9.

Uddin, M.S., Das, G.B., Hossain, A., Rahman, M.A., Mazid, M.A. and Gupta, M.V. (1998). Integrated poultry-fish farming: a way to increase productivity and benefits. In: *Integrated Fish Farming: Proceedings of a Workshop on Integrated Fish Farming Held in Wuxi, Jiangsu Province, People's Republic of China October 11–15, 1994* (eds J.A. Mathias, A.T. Charles and H. Baotong), pp. 65–72. CRC Press LLC, Boca Rotan.

Vaudrey, J.M.P., Getchis, T., Shaw, K., Markow, J., Britton, R. and Kremer, J.N. (2009). Effects of oyster depuration gear on eelgrass (*Zostera Marina L.*) in a low density aquaculture site in Long Island Sound. *Journal of Shellfish Research*, **28**, 243–250.

Williams, M.J. (1997). Aquaculture and sustainable food security in the developing world. In: *Sustainable Aquaculture* (ed. J.E. Bardach), pp. 15–51. John Wiley and Sons, Inc., New York, pp. 15–51.

Wu, H.-J., Lin, Y.-T., Yang, H.-Y., Wan, Y.-H. and Zhang, Q. (1998). Adapting integrated fish farming principles to improve fish yields in Meicuan Reservoir, China. In: *Integrated Fish Farming: Proceedings of a Workshop on Integrated Fish Farming Held in Wuxi, Jiangsu Province, People's Republic of China October 11–15, 1994* (eds J.A. Mathias, A.T. Charles and H. Baotong), pp. 307–324. CRC Press LLC, Boca Rotan.

Yuan, X. (2007). Economics of aquaculture feeding practices: China. In: *Economics of Aquaculture Feeding Practices in Selected Asian Countries* (ed. M.R. Hasan), pp. 65–97. FAO Fisheries Technical Paper No. 505. FAO, Rome.

Zhu, X.-B., Fang, G.-S., Jiang, X.-Y., Li, J.-L. and Mei, Z.-P. (1998). The nature and function of detritus in fish pond ecosystem. In: *Integrated Fish Farming: Proceedings of a Workshop on Integrated Fish Farming Held in Wuxi, Jiangsu Province, People's Republic of China October 11–15, 1994* (eds J.A. Mathias, A.T. Charles and H. Baotong), pp. 147–156. CRC Press LLC, Boca Rotan.

6

Urban Landscapes and Ecosystem Services

Jürgen Breuste,[1] Dagmar Haase[2] and Thomas Elmqvist[3]

[1] Department of Geography/Geology, University of Salzburg, Austria
[2] Institute of Geography, Humboldt University, Berlin, and Helmholtz Centre for Environmental Research GmbH–UFZ, Leipzig, Germany
[3] Department of Systems Ecology and Stockholm Resilience Centre, Stockholm University, Sweden

Abstract

Ecosystem services include all ecosystem functions and processes people and society benefit from in economic terms or related to their quality of life. These range from water and climate regulation, over biodiversity and pollination, to aesthetic and recreational services.

The role of cities in maintaining biodiversity for functional ecosystems is becoming an important topic on the global agenda. In particular urban green spaces – that is forests, trees, parks, allotments or cemeteries – provide a whole range of ecosystem services for the residents of a city. Recreation and climate moderation are highly valued ecosystem services. An increase of built-up land by urban sprawl and densification in the inner parts of a city reduces the much-needed ecosystem services.

Growing urban landscapes

The process of urbanization

Urbanization is a global multidimensional process that is manifest through rapidly changing human population densities and changing land cover. The growth of cities is due to a combination of four forces: natural growth, rural to urban migration, massive migration due to extreme events and redefinitions of administrative boundaries (UN Habitat, 2011). Because urbanization is accelerating, the growth

Ecosystem Services in Agricultural and Urban Landscapes, First Edition. Edited by Steve Wratten, Harpinder Sandhu, Ross Cullen and Robert Costanza.
© 2013 John Wiley & Sons, Ltd. Published 2013 by John Wiley & Sons, Ltd.

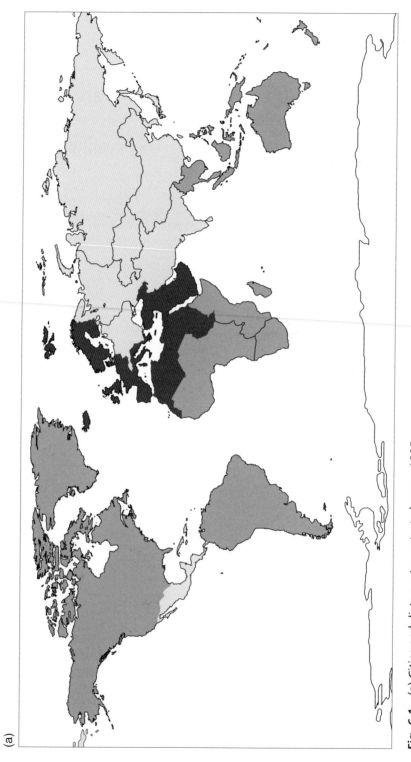

Fig. 6.1 (a) Cities and distance to protected areas, 1995.

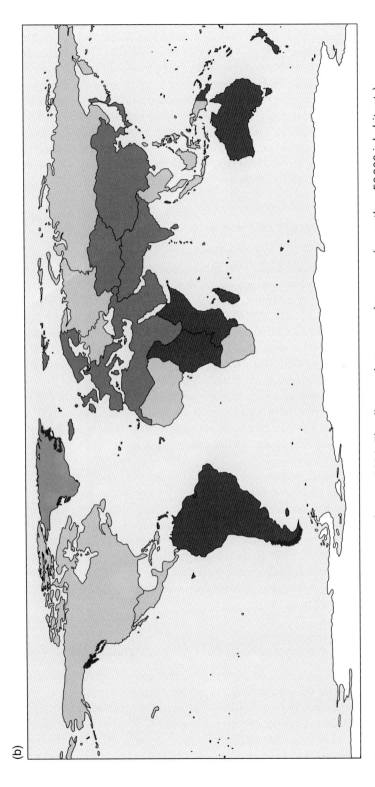

Fig. 6.1 (b) Cities and distance to protected areas, 2030. The distance between urban areas (more than 50 000 inhabitants) and protected areas (PA) in 1995 calculated per region and the projected pattern for 2030 (based on data from McDonald et al., 2008). Black, 25% of PA <15 km; light grey, 25% of PA <25 km; grey, all PA >25 km.

(b)

of cities is leading to the formation of large urban landscapes, particularly in developing countries (Seto et al., 2011). Urban landscape is here defined as an area with human agglomerations with more than 50% of the surface built, surrounded by other areas with 30–50% built, and an overall population density of more than 10 individuals per ha. Urbanization is a process operating at multiple scales and therefore factors influencing environmental change in urban landscapes often originate far beyond city, regional or even national boundaries (McGranahan et al., 2005). Fluctuations in global trade, civil unrest in other countries, health pandemics, lack of sanitation and drinking water access, natural disasters, possibly climate change and political decisions are all factors driving social–ecological transformations of the urban landscape.

Urbanization, biodiversity and ecosystems

In 2010, the International Year of Biodiversity, urbanization was viewed as endangering more species and to be more geographically ubiquitous than any other human activity. The role of cities in maintaining biodiversity for functional ecosystems is becoming an important topic on the global agenda (Convention on Biodiversity, 2010). Urban sprawl is rapidly transforming critical habitats of global value, for example in the Atlantic Forest Region of Brazil, the Cape of South Africa and coastal Central America (Mülller et al., 2010). Urbanization is also viewed as a driving force for increased homogenization of fauna and flora (e.g. Grimm et al., 2008). Furthermore, cities are moving closer to protected areas, particularly in Europe and Asia, a topic that needs increased attention from city planners (McDonald et al., 2008). There is a need for municipalities to be deeply engaged in developing strategies for functional coexistence between dynamic cities, their residents and protected areas (Fig. 6.1).

On the other hand, some cities may also be very rich in biodiversity. A remarkable amount of native species diversity is known to exist in and around large cities, such as Rio de Janeiro, Chicago, Istanbul, Singapore, Cape Town and Stockholm (Elmqvist et al., 2008). Furthermore, a rapid rural to urban migration, particularly in Africa, may in some areas result in reduced pressure on land and considerable re-growth and increase of biodiversity. For a European city, Strohbach et al. (2009) found higher bird diversity in richly structured housing districts, with open backyards containing old trees, compared to the rural surroundings in the region.

Urbanization and management of ecosystems – challenges

Mismatches between spatial and temporal scales of ecological processes on the one hand, and social scales of monitoring and decision making on the other hand, have not only limited our understanding of ecosystem processes in urban landscapes, they have also limited the integration of urban ecological knowledge into urban planning. In ecology, there is now a growing understanding that human processes and culture are fundamental for sustainable management of ecosystems, and in urban planning it is becoming increasingly evident that urban management needs to operate at an ecosystem scale rather than within the traditional boundaries of the city. Although studies of ecological patterns and processes in urban areas have shown a rapid increase during the last decade, there are still significant research gaps that constrain our general understanding of the effects of urbanization processes (Elmqvist et al., 2008). Of further significance is that urban landscapes provide important large-scale experimental study sites of the effects of global change on ecosystems because, for example, significant warming and increased nitrogen deposition already are prevalent and because they provide extreme, visible and measurable examples of human domination of ecosystem processes. Urban landscapes may be viewed as numerous large-scale experiments, producing novel types of plant and animal communities and novel types of interactions among species. The understanding of how urban ecosystems work, how they change, and what limits their performance, can add to the understanding of ecosystem change and governance in general in an ever more human-dominated world (Elmqvist et al., 2008).

Urban ecosystem services

What are urban ecosystem services?

Ecosystem services (ES) include all ecosystem functions and processes people and society benefit from in economic terms or related to their quality of life (Costanza et al., 1997; de Groot et al., 2002). If these ecosystem services are both requested and provided in urban areas and cities, we define them as urban ecosystem services (UES; according to Bolund and Hunhammar, 1999). Typically, ecosystem services that humans benefit from range from water and climate regulation functions, over biodiversity and pollination, to aesthetic and recreational services. Since the first theoretical reflections on ES in the 1990s (Daily et al., 1997; Costanza et al., 1997; de Groot et al., 2002), and certainly with the publication of the Millennium Ecosystem Assessment (MEA, 2005) and the TEEB study (2011), it has became clear that humankind depends on nature and ecosystems, their functions and the variety of processes and fluxes. Nevertheless, ES are often used without being associated any (economic, social) value (Norberg, 1999). Not exclusively but most noticeably, in urban regions and cities where the majority of people live (United Nations, 2008) nature and ecosystems are intensively used and appear to be more and more degraded/ destroyed. They develop into a state where they are no longer able to provide any services (MEA, 2005). According to McDonald (2009),UES are provided at different scales within an urban landscape: at the local scale (e.g. temperature regulation by tree shade, water and pollutant filtration at a single soil plot or timber production in a specific tree estate), at the regional or landscape scale (recreation, climate regulation, biodiversity), and at the global scale (carbon mitigation, contribution to the continental or world-wide gene pool and biodiversity as such).

Classification of UES

According to the Millennium Ecosystem Assessment (MEA, 2005), Costanza et al. (1997) and more recently TEEB (2011), we can define four categories of UES:

- provisioning services (food and timber production, water supply, the provision of genetic resources);
- regulating services (regulation of climate extremes such as heavy rainfall and heat waves, floods and diseases, regulation of water flows, treatment and handling of waste);
- cultural services (recreation and tourism, provision of aesthetic features, spiritual requirements); and, finally,
- habitat and supporting services (soil formation and processes, pollination or energy, matter and nutrient fluxes, biodiversity).

UES can be related to the partially complementary concept of urban quality of life (QoL; Santos and Martins, 2007), which also covers the different dimensions of sustainability from an anthropocentric point of view (Schetke et al., 2010). Table 6.1 lists and compares both concepts of UES and QoL following the three dimensions of sustainability.

Land use – basic information on human influence on ecosystem services

Ecosystem services are clearly related to land use and land cover, both generally and specifically in urban areas (Breuste et al., 1996). Land use is influenced both by social and environmental processes and patterns. In turn, land use influences ecological patterns and processes in cities, which lead to changes in ecological conditions and the broader environmental context such as formation of the urban heat island. Changes in ecological conditions may affect human perceptions and attitudes and influence the formulation of policies with an impact on land use (Pauleit and Breuste, 2011).

Urban ecosystem services are closely related to the usage of urban land. Landscape components are currently described with the terms land cover or land use. These two terms are often used interchangeably, but this is incorrect as they represent fundamentally different aspects of the landscape component. The first term, land cover, describes the physical attributes of the space ('existing material elements'), while the second, land use, is related to how this space is being used by humans ('for what?'). The need for clear definitions is particularly important for the relationship with ecosystem services.

The complex term 'land use' covers completely different aspects such as use for open spaces or building use. However, there are general patterns of use which make it possible to classify general land-use types. Land use is a temporally variable procedure and the term expresses the spatial orientation of utilization procedures (Haase and Richter, 1980; cf. Table 6.2).

Table 6.1   Services and indicators of quality of life related to the dimensions of sustainability (authors' listing according to Millennium Ecosystem Assessment, 2005 and Santos and Martins, 2007).

Sustainability dimension	Urban ecosystem service	Quality of life indicator
Ecology	Air filtration Climate regulation Noise reduction Rain water drainage Water supply Waste water treatment Food production	Health (clean air, protection against respiratory diseases, protection against heat and cold death) Safety Drinking water Food
Social sphere	Landscape Recreation Cultural values Sense of identity	Beauty of the environment Recreation and stress reduction Intellectual endowment Communication Place to live
Economy	Provision of land for economic and commercial activities and housing	Accessibility Income

Table 6.2   Examples of subtypes of residential estates in Leipzig, Germany (by using built-up and open space/ vegetation structures).

Subtypes of residential estates	Period
City centre	
Detached curb-close apartment buildings with built-up courtyards	1870–WW I
Terraced curb-close apartment buildings with built-up courtyards	1870–WW I
Detached curb-close apartment buildings with open courtyards	1900–WW I
Terraced curb-close apartment buildings with open courtyards	1900–WW II
Free-standing blocks of flats in rows	since WW I
Large new prefabricated housing estates	since 1960
Detached and semidetached houses	
Villas	
Former village centres	

Urban green – carrier of UES

Types of urban green space

In particular, urban green spaces (UGS) – that is forests, trees, parks, allotments or cemeteries – provide a whole range of ecosystem services for the residents of

Table 6.3 Urban ecosystems generating local and direct services, relevant for Stockholm (from Bolund, P. and Hunhammar, S. (1999) Ecosystem services in urban areas. *Ecological Economics*, **29**, 293–301).

	Street tree	Lawn/ parks	Urban forest	Cultivated land	Wetland	Stream	Lake/ sea
Air filtering	x	x	x	x	x		
Microclimate regulation	x	x	x	x	x	x	x
Noise reduction	x	x	x	x	x		
Rainwater drainage		x	x	x	x		
Sewage treatment					x		
Recreation/ cultural values	x	x	x	x	x	x	x

a city (Table 6.3). Firstly, they help regulating extreme day- and night-time temperatures by shading, evapotranspiration and lower surface emissivity (Chiesura, 2004; Chang et al., 2007; Kottmeier et al., 2007; Priego et al., 2008). Nearly all types of urban open and green spaces provide recreational facilities. Unsealed land helps to regulate surface water flows, enhances infiltration and lowers water travel times, which help to prevent floods and related damages (Haase, 2003). To support this kind of service by unsealed land, urban water management increasingly uses in situ drainage sites (Haase, 2009; Bastian et al., 2012).

Recreation

Perhaps one of the most important, and therefore highly valued, ecosystem services in cities is recreation, which includes the provision of recreation opportunities by natural and seminatural landscapes to urban residents and the need by urban residents to relax. There is a range of studies on analysing and measuring the recreation function or the recreation ecosystem service (e.g. De Vries et al., 2003; Handley et al., 2003; Chiesura, 2004; Li et al., 2005; Jim and Chen, 2006; Mazuoka and Kaplan, 2008; Comber et al., 2008; Kazmierczak and James, 2008).

Fig. 6.2 shows the growing disconnection between residents and quantity and location of available UGS as recreational places using the example city Leipzig, Germany. This is an overall trend in many cities world-wide. This status of disconnection is measurable (Rink, 2005).

The example in Fig. 6.3 shows urban green space (UGS) supply and demand per capita. The overlay of both graphs along the urban-to-rural gradient shows the dissimilarity in their distribution.

Compared to carbon sequestration by urban vegetation, recreation green space supply and tree shade represent local urban ecosystem services in terms of where they are supplied and consumed (McDonald, 2009).

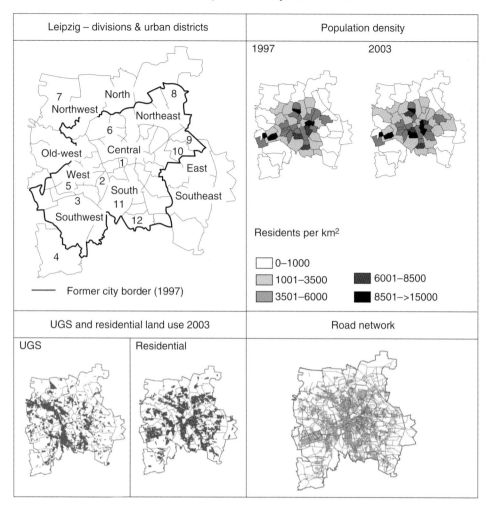

Fig. 6.2 Main features of the case study area that go into the analysis of the demand–supply relation of the recreation ecosystem service for both points in time in 1997 and 2003.

Climate regulation

Starting with the most prominent case of climate change, a range of studies have been carried out to measure the temperature reduction potential and performance of urban green spaces by evapotranspiration and shading (e.g. Jim and Chen, 2006; Gill et al., 2007; Tratalos et al., 2007; Schwarz et al., 2011. A recent study by Vogel and Haase (unpublished) in the city of Leipzig, Germany provided evidence that urban park trees lower the day-time temperature on hot summer days by 2–4 K (Fig. 6.4). Temperature was measured at shaded and non-shaded places in a representative range of urban parks using temperature loggers. Using an urban tree GIS-data layer the shading potential of urban parks P was extrapolated to the whole city (Fig. 6.3):

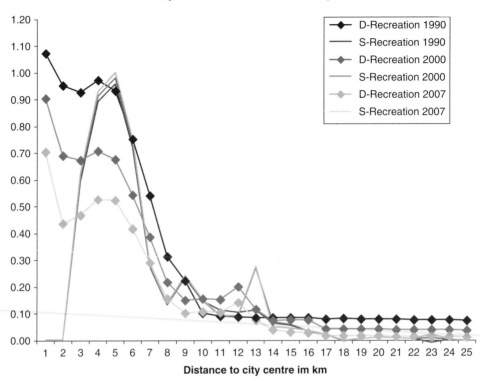

Fig. 6.3 Standardized values illustrating the recreation ecosystem service supply (S) and demand (D) along the urban (0) to rural (25) gradient of Leipzig (Germany). A large undersupply in the inner city areas can be mirrored by a partial oversupply in the floodplain areas at 2–7 km distance from the city centre. In the peri-urban areas demand and supply outweigh each other (based on data from Haase, 2010).

$$\frac{1}{N}\sum_i^n \left(P_{shade}/P_{totalArea} \right)$$
(6.1)

Based on the extrapolated temperature reduction potential values, three explorative scenarios show the relationship between climate regulation potential and land-use change (Fig. 6.5). The first type of scenario assumed: (1) a linear trend of land-use change in the city based on statistical data of the last decade; (2) an enlargement of the green infrastructure due to on-going shrinkage and demolition processes; and (3) the reverse process – reurbanization combined with a transformation of urban green spaces into residential land. In the second type of scenario, the total areas of green space were kept stable but the tree percentage changed: it ranges from (1) the current tree proportion of about 44%, over (2) a complete afforestation of all urban parks, and (3) the most simple form of urban green spaces – lawns – which are considerably cheaper compared to all other types of planted green spaces. Fig. 6.5 impressively shows the effects of these assumed land use changes on the ecosystem service of

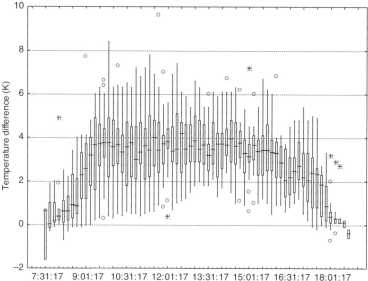

Fig. 6.4 (above) Temperature logging site in an urban park in Leipzig (the red dots show the allocation of the temperature loggers). The box plot below shows the mean temperature lowering potential of the tree shade including the variance of the values expressed by the temperature difference of shaded and non-shaded plots (authors' data from August 2009, max. 36°C air temperature).

temperature reduction by tree shading. We see that both land-use change and a modification of the tree share at prevailing green spaces have an impact on the temperature regulation potential. Most obvious is the positive impact of a green space enhancement following a demolition of inner-urban and peri-urban housing and commercial stock; here the area of an average temperature reduction

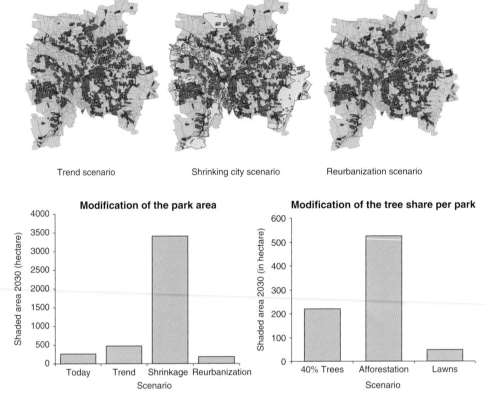

Trend scenario   Shrinking city scenario   Reurbanization scenario

Fig. 6.5 Scenarios of the cumulative temperature reduction potential by tree shading (upper sequence of maps). Below, the total shaded area for (left) the three scenarios 'trend', 'shrinkage' and 'reurbanization' compared to the today's situation, as well as (right) the scenarios '40% trees', 'afforestation' and 'lawns' (at constant park area) is given.

potential of 2–4 K during heat waves multiplies more than tenfold. Vice versa, an increase of built-up land in the inner parts of a city decreases the ecosystem service. Lawns – this also became very clear – are by far the form of urban green space development with the least impact in terms of climate regulation (Fig. 6.5).

Biodiversity

Urban green and natural areas are the most important habitat for plants and animals in the cities. Beside purely aesthetical functions (Priego et al., 2008; Qureshi and Breuste, 2010) the improvement of physical health and nature experience becomes increasingly important (Bolund and Hunhammar, 1999; Chiesura, 2004; Yli-Pelkonen and Niemelä, 2005). These areas provide contact to different common and typically urban but also rare species for urban residents. Often this is for them the only possibility to come in contact with nature. The wide range of nature coexisting in cities offers different habitat/ nature experiences and is in its biodiversity a value to be protected in every city. The vegetation cover (as part of the surface cover) is a component of almost all urban land use types (urban structural units). Some units are dominated by a designed vegetation cover (e.g. parks and allotments), in others the vegetation cover is an additional decorative element (e.g. residential areas). Another type of vegetation cover establishes spontaneously after finishing or interrupting utilization processes for a longer period of time (derelict land).

In Central Europe, cities grew and still grow into cultural (agricultural and only partly forested) landscape. The vegetation cover of open spaces within urban areas ranges from vegetation remnants of the original natural landscape (mainly woods and wetlands) and vegetation of the cultural landscapes formed by agriculture (e. g. meadows and arable land) over ornamental, horticultural and designed urban vegetation spaces (parks and gardens), to spontaneous urban vegetation (brown fields and derelict land). These four main groups of vegetation cover are results of different land use forms (functions) and of different intensities of utilization and maintenance (Table 6.4).

Carbon mitigation

In terms of carbon mitigation, urban trees and urban soils contribute to the carbon uptake and thus to a partial decrease of the urban ecological footprint. Nowak and Crane (2002) estimate that urban trees (and forests) thus are able to balance about 1–2% of urban carbon emissions (Fig. 6.6). While urbanization is increasing globally, with more and more people living in cities, many industrialized cities are losing population – they are shrinking. Concerning carbon storage, shrinking cities are of particular interest: they have a high potential for urban reconstruction and associated new green space as brownfields are abundant and development pressure is low. This is the case in Leipzig, Germany. Fig. 6.6 presents the example of the carbon uptake potential of an urban restructuring project in Leipzig, which illustrates the climate regulation potential of urban restructuring measures in cities. In our example, carbon sequestration occurs by tree growth and was contrasted with all related carbon sources, for example maintenance emissions.

Rapid growth of soil sealing – destruction of UES and its avoidance

Soil sealing is a form of land cover that becomes more prominent with increasing urban influence. It is the main destructor of ecosystem services in urban areas. Soil sealing is the process of removal of the vegetation cover of soils and replacing it with less-permeable materials (bitumen, concrete, stone pavements, etc.) to create building grounds or to build sidewalks, plazas, streets and roads (Breuste, 2009).

Table 6.4 Types of urban vegetation structures – influenced or created by urban land use.

Vegetation group	Vegetation structure type	Main urban ecosystem services and utilization	Main potential functions
(A) Vegetation remnants of the original natural landscape	Woods and forests Wetlands	Recreation, biodiversity Biodiversity	Timber production Nature experience
(B) Vegetation of the cultural landscapes formed by agriculture	Meadows, pastures Drifts, dry grasslands Arable land	Agriculture Agriculture Agriculture	Recreation, biodiversity Recreation, biodiversity
(C) Ornamental, horticultural and designed urban vegetation spaces	Decorative green (flower beds, small lawn patches, bushes, hedges, etc.)	Decoration	Recreation, biodiversity
	Accompanied green along traffic lines or as addition to fill up the space between apartment blocks	Decoration	Recreation, biodiversity
	Gardens/ parks	Recreation, decoration	Biodiversity
	Allotment gardens (territorially organized in allotment garden estates)	Recreation	Biodiversity
	Urban trees	Decoration	Biodiversity
(D) Spontaneous urban vegetation (areas)	Spontaneous herbaceous vegetation	None	Biodiversity, nature experience, recreation
	Spontaneous bush vegetation	None	Biodiversity, nature experience, recreation
	Spontaneous preforest vegetation	None	Biodiversity, nature experience, recreation

Sources of data: Breuste, according to Arbeitsgruppe Methodik der Biotopkartierung im besiedelten Bereich (1993), kowarik, 1992.

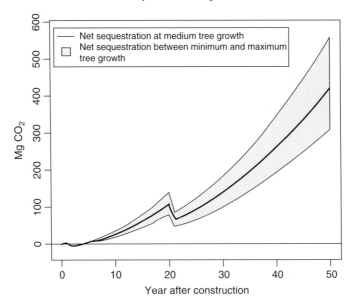

Fig. 6.6 Carbon sequestration of the reconstruction project in 50 years' lifetime. Emissions from construction and management are balanced against sequestration from tree growth. Tree growth is modelled for a range of growth rates. After 5 years of tree growth the balance becomes positive. After 20 years, 182 of the initially 461 trees are thinned out, hence the kink in the line. (From Strohbach MW, Arnold E, Haase D 2012. The carbon mitigation potential of urban restructuring – a life cycle analysis of green space development. *Landscape and Urban Planning*, **104**, 220–229).

The changes to the ecosystem services can be grouped into three major categories: changes to the physical soil structure and the water regime, changes to the microclimatic conditions (temperature, humidity, periodicity of snow cover) and changes to the surfaces available for use by plants and animals (Table 6.5).

Reducing the impact of the sealed surfaces by increasing their water permeability and providing additional infiltration opportunities are goals that require a complex approach to the management of ecosystem services. To achieve these goals, there is a need to improve the monitoring methodology (remote sensing, geographical information systems) and to steer the growth of soil sealing. The best results can be achieved by avoiding soil sealing as much as possible (Breuste et al., 1996; Münchow and Schramm, 1997; Breuste, 2007, 2009, 2012).

Climate change – challenges for UES

While climate change is a profound global issue, it is also a deeply local urban issue, because this is the scale at which most of the immediate impacts are manifest and at which most adaptation actions to cope with climate change are needed

Table 6.5 Destruction and reduction of ecosystem services by soil sealing.

Soil and water regime (by 'loss' of the vegetation cover and physical change of the soil surface and the upper soil layer)	Partial or complete removal of the upper soil layer
	Decreased infiltration of precipitation water into the soil and thus reduced ground-water renewal
	Increased evaporation, and accelerated rates of storm water run-off
	More frequent high tides in drains and water streams with heavy rain and thaw
Urban climate (by 'loss' of the vegetation cover and thermal and energetic effects of the new technical surfaces)	Increased air temperatures by increased thermal capacity and thermal conductivity of the sealing materials
	Increased particulates, and thus more frequent precipitation events
	Lower volume and shorter periods of snow cover
	Reduced humidity in temperate regions
Biodiversity (by destruction of the vegetation cover and change of the local ecological conditions, intensive use by trampling and driving on)	Reduced, usually minimal, colonization opportunities for plants
	Lower oxygen and water supply for soil fauna and decreased exchange of matter and gasses between the soil and the near-surface air layer
	Depletion of the native flora
	Loss of levels of the food pyramid
	Loss of habitat
	Increasing isolation of populations (Breuste, 2009)

(Kates and Wilbanks, 2003). Despite the fact that the world is increasingly urban, the ways in which cities influence, and are influenced by, climate change have been considerably less explored than other areas of research on global warming (Wilbanks et al., 2007). In Table 6.6, projected impacts on urban ecosystems and urban areas of changes in extreme weather and climate events are illustrated (IPCC, 2007).

In light of these projected changes, there is an urgent need to radically rethink conventional urban land use planning and focus on innovative ways of adapting to a new urban landscape. Urban ecosystems must be used for reducing climate change risks, both in the current built-up areas and in the future urban landscape. However, currently this topic represents a substantial knowledge gap that needs to be bridged (MEA, 2005; World Bank, 2009). Below we discuss some climate-change-related stressors in an urban context and how different variables can be managed by improved attention to urban land use and to effects on urban ecosystem services.

Increase in temperature

Projected increases in temperatures are likely to have large-scale effects on the distribution of organisms, particularly at high latitudes (IPCC, 2007). Because cities are the ports of introduction of most exotic species that are dispersed from one continent to another, we may expect large changes in species composition in

Table 6.6   Projected impacts of changes in extreme weather and climate events on urban ecosystems and urban areas.

Climate phenomena and their likelihood	Projected impacts on urban ecosystems and urban regions
Increase in temperature and more frequent hot days and nights, warm spells and heat waves	Changes in species composition, invasion of exotic species
	Increased demand for cooling
	Declining air quality
Very likely to certain	Heat and respiratory stresses
Increased frequency of heavy precipitation events	Changes in species composition in urban ecosystems
Very likely	Disruption of settlements, commerce and transport, loss of property due to flooding
Increased frequency of drought	Loss of drought-intolerant species
Likely	Water shortages for households, industries and services
Increase in storm activity	Likely loss of late successional habitats and large trees
Likely	Damage of property, disruption of water supply and services
Increase in extreme high sea-level	Changes in species composition in urban ecosystems
Likely	Damage of property in coastal cities, cost of coastal protection versus relocation, decreased freshwater availability due to salt-water intrusion

urban ecosystems with multiple effects on ecosystem services. Extreme heat waves may be typical in many cities in the future, which will create new health hazards and cause disruption to public health services, leading to for example increased disease incidence.

Precipitation

According to the IPCC (2007) run-off and water availability are projected to increase by 10–20% by 2100 in some areas and to decrease by 10–30% in areas that are at present water-stressed. Increases in frequency and severity of floods and droughts, as well as declines in water quantities stored in glaciers and snow cover, are expected. Flooding is the most damaging natural disaster world-wide, and is expected to grow during the coming decades because of both climate change and shifting land uses, such as filling of wetlands and expansion of impervious surfaces, which leads to more rapid precipitation run-off into rivers (Opperman et al., 2009). Ecosystems, particularly forests, wetlands and flood-plains, represent important buffering systems for reducing peaks in water flows and also in water retention and purification (McGranahan et al., 2005). However, we are currently witnessing a collapse in the buffering capacity of the hydrological system in many watersheds in urban landscapes and this will increase the impact of floods and aggravate health risks. Increased resilience of water management systems through flood-plain reconnection is, however, a promising example of ecosystem-based adaptation to climate change.

Sea level rise

According to the IPCC (2007), coastal areas are projected to be 'experiencing the adverse consequences of hazards related to climate and sea level (very high confidence) … exposed to increasing risks, including coastal erosion, over coming decades due to climate change and sea-level rise (very high confidence)' (Chapter 6, p. 317). Direct effects of sea level rise include increased storm flooding and damage, inundations, coastal erosion, increased salinity in estuaries and coastal aquifers, rising coastal water tables and obstructed drainage. These direct impacts have knock-on implications for social and institutional systems.

A large proportion of the world's population lives along the coasts and will be at risk as a result of the combined effects of sea level rise and increased frequency and intensity of storms. The same populations have placed considerable pressure on the integrity of coastal and estuarine ecosystems. Because ecosystems such as wetlands, mangrove swamps and coral reefs form natural protections for coastal areas, changes to or loss of these ecosystems will compound the dangers faced by urban coastal areas. Coastal land use and governance structures influence the degree to which this protective role of ecosystems is realized and enhanced or not. In many cases such ecosystems have been destroyed by urban growth even before the issue of global warming was raised in the international agenda (Nicholls and Wong, 2007). The effects of climate change threaten to compound this damage and to reduce these natural protections still further.

UES in urban landscape planning

Not particularly related to UES but in terms of more general environmental, climate change and quality of life concerns, new integrative concepts of urban planning and urban governance are under discussion such as diverse local agendas (Ravetz, 2000; Leser, 2008), the approach of green and blue services (Westerink et al., 2002), or sustainable land-use governance (Ravetz, 2000). Integrative planning thereby means the removal of sectorial thinking and actions in favour of more holistic concepts of regional and local regeneration and adaptation (Ravetz, 2000). Accordingly, expert-driven formal planning will be enhanced and accompanied by participatory processes to ensure a better integration within a city. Not exclusively but most notably in shrinking cities, concepts of green and blue services provide potential to move from a simple 'land-use view' of green and water areas in cities towards a valuation of ecosystem processes and spatial potentials of each piece of land – that is urban ecosystem services (Lorance Rall and Haase, 2011; Bastian et al., 2012). On the basis of

	Old villas	**City centre**

Usage	Residential	Residential, commercial, offices
Built-up structure (type)	Single houses	Compact building blocks
Built-up density	20–30%	More than 70%
Structure of open spaces	Extended open spaces	Small open spaces in courtyards, some squares, streets
Ratio of vegetation and grove	High degree of vegetation, especially tree cover	Mostly no vegetation
Degree of soil sealing	<40%	>90%

Fig. 6.7 The urban structure types of old villa areas and the city centre in Leipzig represent ecologically very different land use forms (based on data from Breuste, 2009).

land-use types (Fig. 6.7), spatial ecological units, urban landscape units and urban structure units of physiognomically homogeneous structure have been developed to balance urban ecosystem services during planning (e.g. Breuste, 1985, 2009; Pauleit and Breuste, 2011).

References

Arbeitsgruppe Methodik der Biotopkartierung im besiedelten Bereich (1993). Flächendeckende Biotopkartierung im besiedelten Bereich als Grundlage einer am Naturschutz orientierten Planung: Programm für die Bestandsaufnahme, Gliederung und Bewertung des besiedelten Bereichs und dessen Randzonen: Überarbeitete Fassung 1993. *Natur und Landschaft*, **68**, 491–526.

Bastian, O., Haase, D. and Grunewald, K. (2012). Ecosystem properties, potentials and services – the EPPS conceptual framework and an urban application example. *Ecological Indicators*, **21**, 7–16.

Bolund, P. and Hunhammar, S. (1999). Ecosystem services in urban areas. *Ecological Economics*, **29**, 293–301.

Breuste, J. (1985). Methodische Aspekte der Analyse und Bewertung der urbanen Landschaftsstruktur. *Proceedings of the VII. International Symposium on Problems of Landscape Ecological Research*. Pezinok (CSSR), Vol. 1, Panel Bo 1, pp. 1–10.

Breuste, J. (ed.) (1996). *Stadtökologie und Stadtentwicklung: Das Beispiel Leipzig*. Analytica, Berlin.

Breuste, J. (2007). Urban soil sealing – key indicator for urban ecological functionality and ecological planning. In: *25 Years of Landscape Ecology: Scientific Principles in*

Practice, Proceedings of the 7th IALE World Congress, Part 1, pp. 197–198. Wageningen, the Netherlands.

Breuste, J. (2009). Structural analysis of urban landscape for landscape management in German cities. In: *Ecology of Cities and Towns: A Comparative Approach* (eds M. McDonnell, A. Hahs and J. Breuste), pp. 355–379. Cambridge University Press, New York.

Breuste, J. (2012). Soil sealing in German cities – forty years investigation. In: *Four Dimensions of the Landscape* (ed. J. Lechnio). Warsaw. *The Problems of Landscape Ecology*, Special Issue **30**, 45–52.

Breuste, J., Keidel, T., Meinel, G., Münchow, B., Netzband, M. and Schramm, M. (1996). Erfassung und Bewertung des Versiegelungsgrades befestigter Flächen. Leipzig (UFZ–Bericht 12/1996).

Chang, C.R., Li, W.H. and Chang, S.D. (2007). A preliminary study on the local cool-island intensity of Taipei city parks. *Landscape and Urban Planning*, 80, 386–395.

Chiesura, A. (2004). The role of urban parks for the sustainable city. *Landscape and Urban Planning*, **68**, 29–138.

Comber, A., Brunsdon, C. and Green, E. (2008). Using a GIS-based network analysis to determine urban green space accessibility for different ethnic and religious groups. *Landscape and Urban Planning*, **86**, 103–114.

Convention on Biodiversity (2010). *Plan of Action on Cities, Local Authorities and Biodiversity 2011–2020*. Available at: http://www.cbd.int (accessed August 2012).

Costanza, R., d'Arge, R., de Groot, R., et al. (1997). The value of the world's ecosystem services and natural capital. *Nature*, *387*, 253–260.

Daily, G., Reichert, J.S. and Myers, J.P. (1997). *Nature's Services: Societal Dependence on Natural Ecosystems*. Island Press, Washington, DC.

de Groot, R., Wilson, M.A. and Boumans, R.M.J. (2002). A typology for the classification, description and valuation of ecosystem functions, goods and services. *Ecological Economics*, **41**, 393–408.

De Vries, S., Verheij, R.A., Groenewegen, P.P. and Spreeuwenberg, P. (2003). Natural environments – healthy environments? An exploratory analysis of the relationship between greenspace and health. *Environment and Planning*, A35, 1717–1731.

Elmqvist, T., Alfsen, C. and Colding, J. (2008). Urban systems. In: *Ecosystems, Encyclopedia of Ecology* (eds S.E. Jørgensen and B.D. Fath), Vol. 5, pp. 3665–3672. Elsevier.

Gill, S., Handley, J.F., Ennos, A.R. and Pauleit, S. (2007). Adapting cities for climate change: the role of the green infrastructure. *Built Environment*, 33, 115–133.

Grimm, N.B., Faeth, S.H., Golubiewski, N.E. et al. (2008). Global change and the ecology of cities. *Science*, **319**, 756–760.

Haase, D. (2003). Holocene floodplains and their distribution in urban areas: functionality indicators for their retention potentials. *Landscape and Urban Planning*, 66, 5–18.

Haase, D. (2009). Effects of urbanisation on the water balance – a long-term trajectory. *Environment Impact Assessment Review*, **29**, 211–219.

Haase, D. (2010). Trade-offs between ecosystem services under conditions of changing land use: the urban perspective. *Global Land Project Workshop on Representation of Ecosystem Services in the Modelling of Land Systems*. Aberdeen, March 2010.

Haase, G. and Richter, H. (1980). Entwicklungstendenzen und Aufgabenstellungen in der Landschaftsforschung der DDR. *Geograficky Casopis*, 4, 231–247.

Handley, J., Pauleit, S., Slinn, P. et al. (2003). Providing accessible natural green space in towns and cities: a practical guide to assessing the resource and implementing local standards for provision. Available at: http://www.english-nature.org.uk/pubs/publication/PDF/Accessgreenspace.pdf (accessed August 2012).

IPCC (2007). *Climate Change 2007: Synthesis Report. Contribution of Working Groups I, II and III to the Fourth Assessment Report of the Intergovernmental Panel on Climate Change*. [Core Writing Team eds Pachauri, R.K. and Reisinger, A.]. IPCC, Geneva.

Jim, C.Y. and Chen, W.Y. (2006). Recreation-amenity use and contingent valuation of urban green spaces in Guanzhou, China. *Landscape and Urban Planning*, 75, 81–96.

Kates, R.W. and Wilbanks, T.J. (2003). Making the global local: Responding to climate change concerns from the ground up. *Environment*, **45**, 12–23.

Kazmierczak, A.E. and James, P. (2008). Urban green spaces: natural and accessible? In: *International Conference Urban Green Spaces – a Key for Sustainable Cities Conference Reader* (eds C. Smaniotto Costa, J. Mathey, B. Edlich and J. Hoyer), 17–18 April 2008, Sofia, pp. 131–134.

Kottmeier, C., Biegert, C. and Corsmeier, U. (2007). Effects of urban land use on surface temperature in Berlin: case study. *Journal of Urban Planning and Development*, **133**, 128–137.

Kowarik, I. (1992). Das Besondere der städtischen Flora und Vegetation. In: *Natur in der Stadt – der Beitrag der Landespflege zur Stadtentwicklung*. Schriftenreihe des Deutschen Rates für Landespflege, H. 61, pp. 33–47.

Leser, H. (2008). *Stadtökologie in Stichworten*, 2nd edn. Borntraeger, Berlin, Stuttgart.

Li, F., Wang, R., Paulussen, J. and Liu, X. (2005). Comprehensive concept planning of urban greening based on ecological principles: a case study from Beijing, China. *Landscape and Urban Planning*, **72**, 325–336.

Lorance Rall, E.D. and Haase, D. (2011). Creative intervention in a dynamic city: a sustainability assessment of an interim use strategy for brownfields in Leipzig, Germany. *Landscape and Urban Planning*, **100**, 189–201.

Mazuoka, R.H. and Kaplan, R. (2008). People needs in the urban landscape: Analysis of landscape and urban planning contributions. *Landscape and Urban Planning*, 84, 7–19.

McDonald, R. (2009). Ecosystem service demand and supply along the urban-to-rural gradient. *Journal of Conservation Planning*, **5**, 1–14.

McDonald, R.I., Kareiva, P. and Forman, R.T.T. (2008). The implications of current and future urbanization for global protected areas and biodiversity conservation. *Biological Conservation*, **141**, 1695–1703.

McGranahan, G., Marcotullio, P., Xuemei, B. et al. (2005). *Urban Systems*, Vol. 27, *Ecosystems and Human Well-being: Current State and Trends*, pp. 795–825. Island Press, Washington, DC. http://www.maweb.org.

MEA (2005). Millennium Ecosystem Assessment. http://www.millenniumassessment.org/en/.

Müller, N., Werner, P. and Kelcey, J.G. (eds) (2010). *Urban Biodiversity and Design*. Wiley-Blackwell, Oxford.

Münchow, B. and Schramm, M. (1998). Permeable pavements – an appropriate method to reduce stormwater flow in urban sewer systems? In: *Urban Ecology* (eds J. Breuste, H. Feldmann and O. Uhlmann), pp. 183–186. Springer, Leipzig.

Nicholls, R.J. and Wong, P.P. (2007). Chapter 6: Coastal systems and low-lying areas. In: *IPCC WGII Fourth Assessment Report*, pp. 316–357. Cambridge University Press.

Norberg, J. (1999). Linking nature's services to ecosystems: some general ecological concepts. *Ecological Economics*, **29**, 183–202.

Nowak, D.J. and Crane, D.E. (2002). Carbon storage and sequestration by urban trees in the USA. *Environmental Pollution*, **116**, 381–389.

Opperman, J., Galloway, G., Fargione, J., Mount, J.F., Richter, B.D. and Secchi, S. (2009). Sustainable floodplains through large-scale reconnection to rivers. *Science*, **326**, 1487–1488.

Pauleit, S. and Breuste, J. (2011). Land use and surface cover as urban ecological indicators. In: *Urban Ecology, Patterns, Processes, and Applications* (eds J. Niemelä, J. Breuste, T. Elmqvist, G. Guntenspergen, P. James and N. McIntryre), pp. 19–30. Oxford University Press, Oxford.

Priego, C., Breuste, J. and Rojas, J. (2008). Perception and value of nature in urban landscapes: a comparative analysis of cities in Germany, Chile and Spain. *Landscape Online*, 7, 1–22.

Qureshi, S. and Breuste, J. (2010). Prospects of biodiversity in the mega-city of Karachi, Pakistan: potentials, constraints and implications. In: *Urban Biodiversity and Design – Implementing the Convention on Biological Diversity in Towns and Cities* (eds N. Müller, P. Werner and J. Kelcey), pp. 497–517. Wiley-Blackwell, Oxford.

Ravetz, J. (2000). *City Region 2020: Integrated Planning for a Sustainable Environment.* Earthscan, London.

Rink, D. (2005). Surrogate nature or wilderness? Social perceptions and notions of nature in an urban context. In: *Wild Urban Woodlands: New Perspectives for Urban Forestry* (eds I. Kowarik and S. Körner), pp. 67–80. Springer, Berlin.

Santos, L.D. and Martins, I. (2007). Monitoring urban quality of life – the Porto experience. *Social Indicators Research*, **80**, 411–425.

Schetke, S., Haase, D. and Breuste, J. (2010). Green space functionality under conditions of uneven urban land use development. *Land Use Science*, **5**, 143–158.

Schwarz, N., Bauer, A. and Haase, D. (2011). Assessing climate impacts of planning policies – an estimation for the urban region of Leipzig (Germany). *Environmental Impact Assessment Review*, **31**, 97–111.

Seto, K.C., Fragkias, M., Güneralp, B. and Reilly, M.K. (2011). A meta-analysis of global urban land expansion. *PLoS ONE*, **6**, e23777.

Strohbach, M.W., Arnold, E. and Haase, D. (2012). The carbon mitigation potential of urban restructuring – a life cycle analysis of green space development. *Landscape and Urban Planning*, **104**, 220–229.

Strohbach, M., Haase, D. and Kabisch, N. (2009). Birds and the city – urban biodiversity, land-use and socioeconomics. *Ecology and Society*, **14**, 31.

TEEB (2011). *The Economics of Ecosystem Services and Biodiversity for International and National Policymakers.* Earthscan, London.

Tratalos, J., Fuller, R.A., Warren, P.H., Davies, R.G. and Gaston, K.J. (2007). Urban form, biodiversity potential and ecosystem services. *Landscape And Urban Planning*, **83**, 308–317.

United Nations (2008). Department of Economic and Social Affairs, Population Division. *World Urbanization Prospects: The 2007 Revision.* Available at: http://esa.un.org/unup (accessed August 2012).

UN Habitat (2011). *Urban World: Waiting for a Solution.* Urban World Series Vol. 4. Available at: http://www.unhabitat.org/pmss/listItemDetails.aspx?publicationID=3263 (accessed August 2012).

Westerink, J., Buizer, M. and Santiago Ramos, J. (2002). *European Lessons for Green and Blue Services in The Netherlands.* Working Paper for Governance and Strategic Planning Scenarios. Available at: www.plurel.net (accessed August 2012).

Wilbanks, T.J., Ensminger, J.T. and Rajan, C.K. (2007). Climate change vulnerabilities and responses in a developing country city: lessons from Cochin, India. *Environment*, **49**, 22–33.

World Bank (2009). *Development Report: Reshaping Economic Geography.*

Yli-Pelkonen, V. and Niemelä, J. (2005). Linking ecological and social systems in cities: urban planning in Finland as a case. *Biodiversity and Conservation*, **14**, 1947–1967.

Part C

Measuring and Monitoring Ecosystem Services at Multiple Levels

7

Scale-dependent Ecosystem Service

Yangjian Zhang,[1] Claus Holzapfel[2] and Xiaoyong Yuan[1]

[1] Institute of Geographic Sciences and Natural Resources Research, Chinese Academy of Sciences, Beijing, China
[2] Department of Biological Sciences, Rutgers University, Newark, USA

Abstract

The scale-dependent feature of ecosystem services is embodied in the scale dependency of ecosystem provider, ecosystem beneficiary, ecosystem service measurement and ecosystem service management. This study discusses each scale-dependent feature of ecosystem services, and two typical case studies are presented to illustrate the scale dependency of ecosystem service. One case deals with a park in one of the world's largest and most developed metropolitan area (New York), which represents local and regional ecosystem services of green space in an urbanized area. The other case covers the Tibet plateau, which represents a nature-dominated ecosystem that provides ecosystem services with both regional and global significance. Such hierarchically structured ecosystem services underline the importance of understanding ecosystem service in an integrated and comprehensive perspective.

Introduction

Ecosystem services, the basis for the existence and development of human society, refer to the benefits human derive, directly or indirectly, from ecosystem processes and functions (Costanza et al., 1997). An ecosystem service value is determined by ecosystem structure and processes at certain temporal and spatial scales. This chapter addresses the scale-dependent features of ecosystem service by first discussing the concepts of spatial and temporal scales, then how the scale determines ecosystem service. Ecosystem service provider, beneficiaries and management are all scale dependent and ecosystem services realized in various scales belong to each corresponding category. The various ecosystem services across scales are illustrated with two case studies ranging from a landscape to regional biome scale.

Scale

Scale refers to the spatial or temporal dimension of an object or process (e.g. size of area or length of time), characterized by both grain and extent (Peterson and Parker, 1998). The grain is the finest level of spatial resolution possible with a given data set (e.g. pixel size for raster data). The extent is the size of the study area or the duration of time under consideration.

The emergence of scale issue in ecological research and its fundamental significance to ecologists originates from the complex hierarchical organization feature of natural processes. Scale is intrinsic to all natural processes and rules (Farina, 1998). It can be classified as measuring scale and intrinsic scale. Measuring scale is the scale humans depend on to perceive the world and gauge the natural process and structure. It belongs to research techniques and develops with technique advancement. The intrinsic scale is the object under study and the ultimate goal of exploring across scales is to reveal the phenomena and rule based on certain scales (Fu et al., 2008). Measuring and intrinsic scales can be expressed as temporal or spatial scales. In describing natural process function, organizational scale is also used. It refers to the ecological hierarchy such as individual, population, community, ecosystem, landscape and biome.

Ecosystem service is scale dependent

Ecosystem services are not provided homogenously across a spatial landscape and they evolve through time. Some services are generated in one location at one time, but the benefit may be realized in a location different from the generation site or/and at another time. For example, the ecosystem service of regulated and extended water provision develops through time by water regulation provided by mountain-top forest that is often remote from the point of service.

The spatial and temporal features of ecosystem service refer to the different services provided by an ecosystem at various temporal and regional scales. In terms of temporal dimension, ecosystem service can be divided into long-term service (decades), seasonal service (year) and short-term service (hours). In terms of spatial dimension, ecosystem service can be considered as global service or regional service. Ecosystem service can be realized at a range of spatial scales, which can be a small wetland or a large forest ecosystem. At a global scale ($>10^6$ km^2), ecosystems provide services in regard to CO_2, N and P cycling and sequestration, and climate regulation (Hufschmidt, 1983). At a biosphere scale (10^4–10^6 km^2), ecosystems provides services of curbing floods, protecting ground water, controlling soil erosion and species habitat. At a landscape scale ($1-10^4$ km^2), ecosystem service can be reflected in decomposing pollutants and providing biodiversity, etc. An ecosystem composed of various species groups (<1 km^2) can serve to decrease noise and dust. In general, ecosystem services at various scales interact in ways that include mutual promotions and mutual constraints. Large-scale and long-term ecosystem services tend to constrain small-scale, periodic ones, while the groups of the latter ones converge to the former one (Clark et al., 1979; Holling, 1992).

The ecosystem service provider is scale dependent. A segment of a population or populations that provide ecosystem service in a given area is conceptualized as a service-providing unit (Luck et al., 2003). The scale of the service-providing unit determines the services output. For example, maintaining pest, weed and disease resistance of crops is provided at the genetic level (Luck et al., 2003); the biological control of crop pests is provided at the population and food-web level (Wilby and Thomas, 2002); water flow regulation service by vegetation is provided at the habitat and community level (Guo et al., 2000).

The ecosystem beneficiary is scale dependent

Since ecosystem service is provided in a scale-dependent pattern (Millennium Ecosystem Assessment, 2005), the corresponding beneficiaries exist across a range of scales as well. The beneficiary of ecosystem service can be classified into a hierarchy of socioeconomic institutions (Becker and Ostrom, 1995; O'Riordan et al., 1998), which ranges from the lowest institutional level, such as individuals and households, to higher level such as communities or municipalities, then to states or provinces, to nation, and the world. Stakeholders at each scale pay attention to the ecosystem service in which they have an interest and their utilization of ecosystem service likewise may vary greatly. For example, local residents value the timber woods of a forest, while state government pays more attention to its value for recreation or culture, and international communities see its value in offsetting global warming. Since ecosystem service and the service beneficiary both exist at a range of scales, participants are likely to step across their corresponding scale boundaries and conflict of interest results. From the standpoint of ecosystem service management, it is necessary to identify the complex ecosystem service structure.

Ecosystem service at certain spatial and temporal scales points to specific beneficiaries. The value of ecosystem service is highly related to the action of the beneficiary. It is perceivable that a service cannot be utilized if, for example, an ecosystem is providing a product for a short period only (e.g. wild berries) and immediate harvesting of this product is not possible.

Ecosystem service measurement is scale dependent

Ecosystem services are often ignored by policy makers since most of them have no direct commercial market values. Calculating ecosystem service in economic metrics might assist in improving the awareness of the public and the policy makers. However, ecosystem service measurement is highly likely to be biased due to a number of constraints.

The value of ecosystem service providers is highly likely to be over estimated since the same provider serves in different, often opposing, ways. To avoid such duplicate calculation, it is necessary to frame the ecosystem service into corresponding spatial and temporal scales. For example, a patch of forest interests local people for its timber value. At a global scale, it serves in reducing CO_2 levels. In this case, timber production service can not be counted at the larger, global scale.

Ecosystem boundary delineation can affect ecosystem service assessment fundamentally. Simply defining ecosystem boundaries based on easily identified physical boundaries, such as a lake or a stream, often is inadequate to address the complexity of natural systems within the question being addressed. However, some ecosystem processes or features coincide with the physical boundaries of certain area. For example, productivity calculation of a lake ecosystem can be simply conducted within the delineated lake boundaries, while the nutrient cycling of the lake ecosystem involves many processes crossing the lake boundaries, such as water flow and precipitation. It is challenging to define ecosystem boundaries since highly mobile organisms and constituents interact at multiple spatial and temporal scales. The scale-dependent features of ecosystem processes determine that the conceptualization of an ecosystem and the scope and validity of questions being asked within that ecosystem entail an appropriate choice of boundaries of an ecosystem (O'Neill et al., 1986).

Ecosystem service value assessment is constrained by limited understanding of ecosystem structures and processes across scales (Naeem, 2009). Ecosystem service identification is the basis for evaluating ecosystem service. The features of ecosystem service varying over temporal and spatial scales should be emphasized. In light of the dependence of ecosystem service on ecosystem processes, understanding ecosystem processes is pivotal for assessing ecosystem service. An ecosystem includes all the organisms living in a particular area, and all the non-living, physical components of the environment with which the organisms interact, such as air, soil, water and sunlight (Odum, 1971). From the standpoint of valuing an ecosystem, an ecosystem can be interpreted as interactions between biological organisms and environment which as a whole can output services at various temporal and spatial scales.

The ecosystem service assessment is a subjective process. The assessment result can change with the distance of the ecosystem to a population centre, the fragmented nature of an ecosystem, the purchasing power of people and the spatial scale (Konarska et al., 2002). Biological productivity capacity of an ecosystem varies with product volume, as well as human's preference, harvesting technique and processing technique.

The organizing function of an ecosystem is complex to analyse since functions vary spatially. For example, forest is effective in conserving water supply, but the conserving efficiency changes with spatial scale. Lessening the flood risk involves only the interests of some specific regions. Some ecosystem services, such as N, P and CO_2 cycling, occur only at global scales and analysis over smaller scale is not entailed.

The process of calculating ecosystem service is constrained by lack of a stand-ardized framework and methods (Post et al., 2007). Data utilized in calculating ecosystem service are often in incompatible scales with the ecosystem service itself and non-standardized methods or data would result in different conclusions (de Groot, 2002). There are two conventionally used ecosystem service calcula-tion paradigms. The first is to extrapolate estimated results of a few habitat types to entire regions or the entire planet (Troy and Wilson, 2006; Turner et al., 2007). The second is to focus on a single service in a small area (Kaiser and Roumasset, 2002; Ricketts et al., 2004). The first paradigm is limited in that spatial heterogeneity within one type of habitat is not considered; the second one fails to include the scope (number of services) and scale (geographic and tempo-ral) which are critical for most policy questions (Nelson et al., 2009). Ecosystem service evaluation related to certain specific ecosystems or nations is inadequate to characterize the ecosystem at the global scale. Evaluation at any scale can benefit from the evaluation at higher or lower hierarchy scale (Millennium Ecosystem Assessment, 2005).

As remotely sensed images are becoming increasingly available for global cov-erage and at a continuous temporal scale, they become an important data source for calculating ecosystem service. These images have the advantage of providing spatially explicit information that is readily accessible, as required in assessing ecosystem services, such as land use and land cover, areal extent of each land use type, etc. As a result, the ecosystem functions, goods and service can be evaluated and reported in a spatially explicit manner. A summary of large-scale ecosystem service entails remotely sensed raster data, whose resolution will affect ecosystem service calculation significantly. The various resolutions of remote sensing data might alter the extent of fragmented land cover or leads to disappearance of certain land-cover types (Turner et al., 1989; Moody and Woodcock, 1994). The extent and the land-cover types normally determined from remotely sensed data can significantly influence the ecosystem service values. When remotely sensed land cover is used as a proxy for ecosystem service, the spatial scale at which the land cover is measured significantly influences measurements of both the ecosys-tem service extent and its valuation (Konarska et al., 2002). Results for individual trees can only be identified with fine-resolution data. On the other hand, fine-resolution data can identify the small coverage area, such as a small lake corner or a narrow river. Then extent of these complex landscape bodies will be expanded by fine-resolution data. Consequently, the related ecosystem services will be increased. For example, NOAA-AVHRR imagery and National Landcover Dataset (NLCD) are the two commonly used remote sensed imageries that have entire US coverage. The prior one has a spatial resolution of 8 km and the latter has a spatial resolution of 30 m. When they were used in calculating total ecosystem service value of the USA, it was found that ecosystem service in all states except New Mexico had higher ecosystem service values when measured in NLCD data than AVHRR data. The total ecosystem service value of the USA measured using fine-resolution data is 198% higher than measured using lower-resolution data (Konarska et al., 2002).

Ecosystem service management decision making is scale dependent

The interests that humans obtain from an ecosystem are highly related to its spatial and temporal scales. Ecosystem management should be in accordance with the characteristics of the ecosystem. The primary ecosystem service can only be realized at certain temporal and spatial scales, which means that ecosystem process and service is constrained to a certain extent and period. Ecosystem valuation results at a global scale are unable to meet the need of the policy making for a nation or a region. Appreciating the scale dependency of ecosystem service is pivotal for determining the interests of different stakeholders and establishing compensation payments to local stakeholders that face opportunity costs of ecosystem conservation. It is critical to make decisions on landscape-level conservation and management plans and ecosystem management at an appropriate institutional scale and implement ecosystem conservation and land-use planning (Tacconi, 2000). Separating ecosystem services into distinct scales is important in allocating interests appropriately to the stakeholders. Examples are determinations of the forest area in a watershed that help maintain clean water downstream, distribution pattern of natural habitat patches that provide pollination and pest control services for crops, effect distances of adjacent land uses that affect the capacity of forest and soil ecosystem to purify water (Houlahan and Findlay, 2004). All these services need to be assessed at their corresponding scales.

Ecosystem service types

Ecosystem service can be broadly classified as operating on local, regional, national or global scales (Kremen, 2005). For example, pest controls in crops using native parasitoids and predators conventionally operate at a local scale, while forest contributions to carbon sequestration function at a global scale. Ecosystem services value can be categorized into four types (de Groot et al., 2002; Hein et al., 2006):

1   Direct-use values are all production services and some cultural services (such as recreation) that human can utilize directly (Pearce and Turner, 1990). Typical examples include the wood timber produced by forest, fruits and water (Balick and Mendelsohn, 1992; Pearce and Moran, 1994). Cultural services can be exemplified as benefits people obtain from actual visits, recreation, cognitive development, relaxation and spiritual reflection (Aldred, 1994).
2   Indirect-use values arise from the positive functions provided by ecosystems that humans can utilize indirectly (Munasinghe and Schwab, 1993). The indirect-use value is commonly related to the regulation service provided to society, such as water conservation, carbon sequestration, erosion and flood control, regulating climate, hydrological and biochemical cycles, earth surface

processes and a variety of biological processes which account for a significant proportion of ecosystem service (Tobias and Mendelsohn, 1991; Chopra, 1993; Smith, 1993).

3   Option values are characterized by the willingness to pay in order to keep the option of using a resource in the future (Pearce and Turner, 1990). Ecosystem service is temporal-scale dependent, which means that actual and potential future services provided by ecosystems need to be considered in the valuation (Maler, 2000). A future resource can be any current ecosystem service value.

4   Non-use values can be considered anthropocentric (such as natural beauty), or ecocentric (e.g. relating to the notion that animal and plant species may have an existence right) (Hargrove, 1989) and are inherent to the ecosystem (Van Koppen, 2000). Non-use values vary with the moral, aesthetic and other cultural perspectives of the stakeholders involved. Kolstad (2000) further divides non-use value into three categories: existence value, altruistic value and bequest value.

Some agents provide services related to several values. For example, water provides materials related to human daily lives, such as freshwater and fishes, etc. In addition, freshwater provides a range of services related to regulating services and cultural services, such as tourism, natural flood control and erosion control (Van Jaarsveld et al., 2005).

Each category has a distinct scale-dependent feature. Production service can be accounted for by quantifying the flows and goods harvested in the ecosystem in a physical unit. The regulation service entails spatially explicit analysis of the biophysical impact of the service on the environment in or surrounding the ecosystem. For example, fire impact and hydrological services of a forest need to be evaluated across scales. Carbon sequestration service is an exception that is not scale dependent.

Ecosystem service studies need to consider scale

The multiscale feature of ecosystem services is becoming more evident in the increasingly interconnected global economy environment. Ecosystem services provided at one location can have important implications in far away places. As environmental effects on ecosystem service may be uncorrelated across scales, studies should be ideally carried out at multiple, nested scales (Sayer and Campbell, 2004). Ecosystem service research conventionally treats an ecosystem as an integral entity, while the spatial heterogeneity within an ecosystem is ignored. Ecosystem services can move out of the ecosystem boundary and generate services in areas beyond the system. For example, water conserved by forest in an upper river area can generate ecosystem service outside the forest. The water leaving the system can be used to generate power and irrigate farmland in a downstream area. One type of ecosystem process can generate various types of ecosystem services. Some types of ecosystems services are realized by certain ecosystem processes in the same spatial range. Some ecosystem services can be accumulated in the process of ecosystem processes being converted into ecosystem services, and the accumulation process might involve the spatial shifting from within the ecosystem to outside. In addition, the spatially heterogeneous structure within an ecosystem entails the spatially explicit information as revealed in the ecosystem service result. The spatially heterogeneous features of ecosystem service, including within and outside the ecosystem entity, underpins the necessity for accurately identifying, quantifying and spatially locating it in achieving the goal of precisely valuing the ecosystem service.

Recently, research evaluating ecosystem service across spatial and temporal scale has been increasingly reported (Holmes et al., 2004; Swift et al., 2004; Van Jaarsveld et al., 2005; Yue et al., 2005; Zhang and Lu, 2010). The study conducted in the 20 000 km² Ruoergai Plateau Marshes in the northeastern fringe of Qinghai-Tibetan Plateau found that the ecosystem service value of gas regulation and water regulation accounts for 49.9% and 45.6% of the total ecosystem service value, respectively. While the other ecosystem service items, including livestock products, waste treatment and recreation, account for only 4.5% of the total (Zhang and Lu, 2010). The dominance of regulation service is related to its strong water-holding capacity and carbon sequestration capacity. The extremely harsh living condition makes the marshes less suitable for producing goods to support human life. Another study assessing ecosystem service across spatial scales in the De Wieden wetland in the Netherlands concluded that goods production service, including reed and fish provision, accounted for 14% of the total ecosystem service value. Recreation accounts for 37% and nature conservation accounts for 49% of the total (Hein et al., 2006).

A city can be developed to provide a balanced proportion of each category of ecosystem service. Shenzhen is a typical city in China that experienced rapid development from a village of hundreds of residents since the opening of China to the world. Now the total land area of Shenzhen is 0.19 million ha and the total population is 10.4 million. In 2004, the woodland, cropland, wetland and built-up land accounted for 31%, 18%, 10% and 43% of the total land, respectively. The ecosystem services of water supply, waste treatment and food and raw material provision at the scale of Shenzhen city accounted for 64% of the total, while the ecosystem services at the province scale, including waste treatment and recreation, accounted for 7% of the total. The ecosystem services related to the global scale, including gas and climate regulation, biodiversity protection and recreation, accounted for 29% of the total (Li et al., 2010).

Case studies

Scale dependence of ecosystem service, as discussed above, is illustrated with two cases that consider different ranges of scales and different types of service. The first is a large, polluted, former brownfield site located in Liberty State Park (New Jersey, USA), which developed into an unmanaged wild area and represents a small-scale island of wildland within an urban, metropolitan landscape. The second is the Qinghai-Tibet Plateau (QTP), China, which represents a case in a natural and wild area at an ecoregion scale.

Liberty State Park Interior

Liberty State Park (LSP) Interior is an approximately 100-ha brownfield wild area located in Jersey City, NJ (40° 42′ 16, 74° 03′ 06). For much of the twentieth century, the site at LSP was used as a rail yard and experienced heavy industrial use while acting as a major hub for New York City. By the late 1960s, the rail yard was abandoned and since then the wild area of LSP has undergone a natural, unaided succession that resulted in a diverse mosaic of plant communities. Today, at first glance, the site appears to be a fairly healthy urban ecosystem consisting of a rather eclectic collection of early and mid successional habitats, including shrublands, pioneer hardwood forested wetlands, emergent marsh and more open forb-dominated old field communities and grasslands (Gallagher et al., 2008, 2011). Because of this variety in habitat and the large area of contiguous open space in the middle of a dense urban environment, LSP supports a diverse fauna and flora (US Army Corps of Engineers, New York District, 2004) and there is little doubt that LSP has an integral role in supporting wildlife in the greater New York City area. Thus this area developed into what we define here as an 'urban wildland', an urban habitat initially created by human impact (e.g. by severe disturbance) that either developed naturally and unaided and/or is now in a state of wild (i.e. has no or little direct, continued human impact).

A research team, consisting chiefly of scientists from Rutgers University, is currently investigating the ecosystem functions and their services of this urban wildland with the ultimate goal of developing a ecosystem service metric typical for and usable in urban areas in general (Hofer et al., 2010) For this, the research concentrates on the following services: (1) islands of biodiversity in a matrix of 'urban desert', oasis effects; (2) bioclimatology: amelioration of urban heat islands and air pollution; (3) wildland vegetation as carbon sinks; (4) ground water: improvement of infiltration and filtering functions of urban wetlands; (5) soil amelioration; (6) spaces where natural processes continue to work, including evolutionary processes; and (7) human interface: place for contact with nature and natural processes.

The ecosystem service of urban green space can be realized mostly in providing islands of biodiversity, ameliorating urban heat island effect and air pollution, sinking carbon and providing places for human contact with nature and natural processes. In many cases, an area of urban green space of the same size as an area of rural land can provide a much larger ecosystem service in such aspects as maintaining biodiversity. By definition, urban areas are characterized by strong human impacts. As such, urban ecosystems are expected to be impoverished in species richness compared to regions where human impact remains relatively low (McKinney, 2002). That urban areas, however, can harbour a relative large number of wild species comes for most people, urban and rural dwellers alike, as quite a surprise. Plant species richness and evenness of plant communities often increase in urban environments as compared to rural areas (Hope et al., 2003; Marzluff, 2005; Grove et al., 2006). This appears to be due to high spatial heterogeneity of urban habitats in combination with introductions of non-native, but urban-adapted species (Grimm et al., 2008). Bird diversity as well can increase due to limited urbanization (Marzluff, 2005). This is mainly the consequence of the opening of homogenous, natural habitats such as forests. It can be shown that the green spaces set aside from overly strong anthropogenic pressures (vacant lands, less frequented and less manicured parts of parks, public rights-of-way, residential yards, etc.) act as biodiversity hotspots that contribute to the ecological functions of urban areas. Besides providing such basic ecosystem service functions, these areas provide the unique opportunity for nature–human contact even in cities (Dunn et al., 2006; McKinney, 2006; Grimm et al., 2008). Such contacts are being increasingly recognized as crucial for the welfare of urban humans, as demonstrated by a recent correlations of health aspects with exposure to natural environments (Mitchell and Popham, 2008).

The wildland of LSP is an example of a species-rich island in a matrix of low diversity within a sea of low diversity typical for the built-up and developed urban matrix. Table 7.1 provides an overview of this biodiversity concentration for a number of plant and animal taxa. Table 7.1 illustrates that the precipitous differences in species richness and therefore biodiversity between wildland and urban matrix is consistent between scales. Regardless of whether one considers small-scale community level (here $1000\,m^2$), metacommunity scale (here 200 ha) or regional scale (state wide scale = $22\,590\,km^2$), wildlands tend to harbour much larger biodiversity. Such scale-independent differential effects when considering intensively human-used patches and landscape with lesser-used sites has been noted before (Savard et al., 2000). Table 7.1 provides an overview of this biodiversity concentration for a number of plant and animal taxa. Currently, a frame-work is being developed to assess a network of wildlands on different scales. New Jersey is an ideal proving ground for such studies as the state has a sizable percentage of vacant, postindustrial sites (Lurie and Wacker, 2009). As such, the work will allow meaningful comparisons with other industrialized region of the world.

Table 7.1 Scale-dependent richness within the urban wildland of Liberty State Park and the surrounding matrix. These numbers are preliminary and unpublished data, provided and assembled by a variety of sources and authors; regional data are estimates. Only regularly occurring species are included.

Taxa	Liberty State Park		Urban matrix in New Jersey		Region: New Jersey, urban, % of state	Region: New Jersey, undeveloped Urban, % of state
	0.1 ha	200 ha	0.1 ha	200 ha	26.3%	1.1%
Vascular plants	20–65	185	6–12	45	210	560
Birds	8–17	87	4–7	21	45	210
Mammals	2–8	11	2–4	6	17	45
Odonata	5–7	12	1–3	6	14	35
Lepidoptera (butterflies)	12–14	25	4–5	11	24	50

Qinghai-Tibet plateau

Qinghai-Tibet plateau (QTP) is the largest and highest plateau in the world. It covers about 2 500 000 km² and hosts about 8 000 000 people. Qinghai-Tibet plateau has a unique feature in terms of ecosystem service since its global and continental ecological services have a much more significant meaning than the goods it provides. QTP ecosystem plays an important role in regulating atmospheric chemical composition because forest and grassland can be a huge sink or source for such atmospheric gas as CO_2 and O_2 etc. The critical role of QTP in regulating climate stems from its extensive areas and high plateau. Due to its existence, the area lying to the east of QTP in mid-latitude China receives more rainfall than other areas in the mid-latitudes of the world. There are 29 182 km² of lakes and 65 548 km² of glaciers. The vegetation, lakes and glaciers set the stage for the critical role of QTP in supplying and regulating water for China and other southern Asian countries.

The ecosystem services that the QTP can provide include: (1) food production and provision of raw materials; (2) the provision of opportunities for recreation and culture; (3) generic resources; (4) waste treatment; (5) soil formation and reserve; (6) water regulation and supply; (7) global climate regulation; and (8) atmospheric gas regulation.

The QTP ecosystem services include the provision of goods in the form of grass for livestock grazing and agricultural products to local residents, which form the livelihood of the people. The QTP ecosystem produces 1 790 000 tons of food supply, 345 950 tons of oil, 438 750 tons of meat and 419 700 tons of milk. The total provision of goods and food amount to about 623×10^8 Chinese yuan, which accounts for only 6.5% of the total ecosystem service values (Table 7.2).

Table 7.2 Qinghai-Tibet Plateau: ecosystem service values for each service item (10^8 yuan year^{-1}) (based on data from Xie et al., 2003)

	Forest	Grassland	Farm land	Wetland	Water body	Desert	Total
Gas regulation	470	500	13	5	0	0	988
Climate regulation	362	562	24	50	12	0	1010
Water regulation and supply	430	500	16	46	526	25	1543
Soil formation and reserving	523	1218	39	5	0.3	17	1802.3
Waste treatment	176	818	44	54	469	8	1569
Genetic resources	438	681	19	7	64	286	1495
Food production	13	187	27	1	3	8	239
Raw materials	349	31	3	0.2	0.3	0	383.5
Recreation and cultural	172	25	0.3	16	112	8	333.3

The unique natural beauty, history and culture of the area attract millions of tourists each year. The opening of Qinghai-Tibet railway greatly improved the transportation conditions to Tibet and the number of tourists visiting Tibet is increasing rapidly each year. In 2002, the recreation and cultural ecosystem service value reached 333.3×10^8 yuan, and this number has been increasing annually. In 2009, the Xizang and Qinghai provinces lying in the Tibet plateau attracted over 12 millions of Chinese and foreign tourists.

The ecosystem services with continental and global significance, including atmospheric gas and global climate regulation, and water regulation, soil formation and genetic resources account for 73% of the total. QTP, as the 'roof of the world' is the largest 'water tower', from which many of major river systems originate. It hosts unique biodiversity in the high plateau area. Due to its unique high altitude, mountainous topography and climatic conditions, there are large number of species that found refuge during the ice ages and a number of new species evolved in situ, all of which contribute to a rich genetic resource. The relatively young geological history and the extreme high plateau climate make soils in the QTP high plateau diverse and unique. They also serve as huge reserves of plant nutrients.

Tibet is a typical case that has regional, continental and global ecosystem service significance and the continental and global ecosystem service might have higher significance than the local one due to its ecological significance to the region and globe. At the intrinsic scale, the interests of the local stakeholders of QTP are related mainly to raw materials provided and food production. At a national scale, the interested stakeholders will consider recreation and cultural, water regulation and supply, and waste treatment. At a continental scale, the gas regulation and climate regulation, soil formation and reserves, and genetic resources functions of QTP play a critical role. In addition, the gas and climate regulation function and the genetic resources function have a global impact considering the wide-ranging effect of QTP.

Conclusions

Ecosystem service is scale dependent as revealed by the scale-dependent ecosystem service provider, scale-dependent beneficiary and management. Here, ecosystem service are exemplified in hierarchical levels as characterized in the various types of ecosystem service provided by a park in a metropolitan New York area and the nature-dominated Tibet plateau. To effectively manage and fully utilize ecosystem service of each ecosystem, we need to understand the scale dependency of ecosystem functions. This work was supported by the "One Hundred Talent Plan" of Chinese Academy of Sciences.

References

Aldred, J. (1994). Existence value, welfare and altruism. *Environmental Values*, 3, 381–402.
Balick, M.J. and Mendelsohn, L. (1992). Assessing the economic value of traditional medians from tropical rain forest. *Conservation Biology*, 6, 128–130.

Becker, C.D. and Ostrom, E. (1995). Human ecology and resource sustainability: the importance of institutional diversity. *Annual Review of Ecology and Systematics*, **26**, 113–133.

Chopra, K. (1993). The value of non-timber forest products: an estimation for tropical deciduous forests in India. *Economic Botany*, **47**, 251–257.

Clark, C.W., Jones, D.D. and Holling, C.S. (1979). Lessons for ecological policy design: a case study of ecosystem management. *Ecological Modelling*, **7**, 22–53.

Constanza, R., d'Arge, R., de Groot, R., et al. (1997). The value of the world's ecosystem services and natural capital. *Nature*, **387**, 253–260.

de Groot, R.S., Wilson, M.A. and Boumans, R.M.J. (2002). Typology for the classification, description, and valuation of ecosystem functions, goods, and services. *Ecological Economics*, **41**, 393–408.

Dunn, R.R., Gavin, M.C., Sanchez, M.C. and Solomon, J.N. (2006). The pigeon paradox: Dependence of global conservation on urban nature. *Conservation Biology*, **20**, 1814–1816.

Farina, A. (1998). *Principles and Methods in Landscape Ecology*, pp. 35–49. Chapman and Hall, London, UK.

Fu, B., Lu, Y. and Chen, L. (2008). The latest progress of landscape ecology in the world. *Acta Ecologica Sinica*, **28**, 798–804.

Gallagher, F.J., Pechmann, I., Bogden, J.D., Grabosky, J. and Weis, P. (2008). Soil metal concentrations and vegetative assemblage structure in an urban brownfield. *Environmental Pollution*, **153**, 351–361.

Gallagher, F.J., Pechmann, I., Holzapfel, C. and Grabosky, J. (2011). Altered vegetative assemblage trajectories within an urban brownfield. *Environmental Pollution*, **159**, 1159–1166.

Grimm, N.B., Faeth, S.H., Golubiewski, N.E., et al. (2008). Global change and the ecology of cities. *Science*, **319**, 756–760.

Grove, J.M., Troy, A.R., O'Neil-Dunne, J.P.M., Burch, W.R., Cadenasso, M.L. and Pickett, S.T.A. (2006). Characterization of households and its implications for the vegetation of urban ecosystems. *Ecosystems*, **9**, 578–597.

Guo, Z.W., Xiao, X.M. and Li, D.M. (2000). An assessment of ecosystem services: water flow regulation and hydroelectric power production. *Ecological Applications*, **10**, 925–936.

Hargrove, E.C. (1989). *Foundations of Environmental Ethics*. Prentice-Hall, Englewood Cliffs.

Hein, L., van Koppen, K., de Groot, R.S. and van Ierland, E.C. (2006). Spatial scales, stakeholders and the valuation of ecosystem services. *Ecological Economics*, **57**, 209–228.

Hofer, C., Gallagher, F.J. and Holzapfel, C. (2010). Metal accumulation and performance of nestlings of passerine bird species at an urban brownfield site. *Environmental Pollution*, **158**, 1207–1213.

Holling, C.S. (1992). Cross-scale morphology, geometry and dynamics of ecosystems. *Ecological Monographs*, **62**, 447–502.

Holmes, T.P., Bergstrom, J.C. and Huszar, E. (2004). Contingent valuation, net marginal benefits, and the scale of riparian ecosystem restoration. *Ecological Economics*, **49**, 19–30.

Hope, D., Gries, C., Zhu, W., et al. (2003). Socioeconomics drive urban plant diversity. *Proceedings of the National Academy of Sciences USA*, **100**, 8788–8792.

Houlahan, J.E. and Findlay, C.S. (2004). Estimating the critical distance at which adjacent land-use degrades wetland water and sediment quality. *Landscape Ecology*, **19**, 677–690.

Hufschmidt, M. (1983). *Environment, Natural Systems and Development, and Economic Valuation Guide*. John Hopkins University Press, London.

Kaiser, B. and Roumasset, J. (2002). Valuing indirect ecosystem services: the case of tropical watersheds. *Environment and Development Economics*, **7**, 701–714.

Kolstad, C.D. (2000). *Environmental Economics*. Oxford University Press, New York, Oxford.

Konarska, K.M., Sutton, P.C. and Castellon, M. (2002). Evaluating scale dependence of ecosystem service valuation: a comparison of NOAA-AVHRR and Landsat TM data-sets. *Ecological Economics*, **41**, 491–507.

Kremen, C. (2005). Managing ecosystem services: what do we need to know about their ecology? *Ecology Letters*, **8**, 468–479.

Li, T., Li, W. and Qian, Z. (2010). Variations in ecosystem service value in response to land use changes in Shenzhen. *Ecological economics*, **69**, 1427–1435.

Luck, G.W., Daily, G.C. and Ehrlich, P.R. (2003). Population diversity and ecosystem services. *Trends in Ecology and Evolution*, **18**, 331–336.

Lurie, M.N. and Wacker P.O. (eds) (2009). *Mapping New Jersey. An Evolving Landscape*. Rutgers University Press, New Brunswick.

Maler, K.G. (2000). Development, ecological resources and their management: a study of complex dynamic systems. *European Economic Review*, **44**, 645–665.

Marzluff, J. (2005). Island biogeography for an urbanizing world: how extinction and colonization may determine biological diversity in human-dominated landscapes. *Urban Ecosystems*, **8**, 157–177.

McKinney, M.L. (2002). Urbanization, biodiversity, and conservation. *BioScience*, **52**, 883–890.

McKinney, M.L. (2006). Urbanization as a major cause of biotic homogenization. *Biological Conservation*, **127**, 247–260.

Millennium Ecosystem Assessment (2005). *Ecosystems and Human Well-being: Biodiversity Synthesis*. World Resources Institute, Washington, DC.

Mitchell, R. and Popham, F. (2008). Effect of exposure to natural environment on health inequalities: an observational population study. *Lancet*, **372**, 1655–1660.

Moody, A. and Woodcock, C.E. (1994). Scale-dependent errors in the estimation of land-cover proportions: implications for global land-cover datasets. *Photogrammetric Engineering and Remote Sensing*, **60**, 585–594.

Munasinghe, M. and Schwab, A. (1993). *Environmental Economics and Natural Resource Management in Developing Countries*. World Bank, Washington, DC.

Naeem, S. (2009). Biodiversity, ecosystem functioning, and ecosystem services. In: *The Princeton Guide to Ecology* (ed. S.A. Levin), pp. 584–590. Princeton University Press, Princeton.

Nelson, E., Mendoza, G., Regetz, J., Polasky, S., Tallis, H. and Cameron, D.R. (2009). Modeling multiple ecosystem services, biodiversity conservation, commodity production, and tradeoffs at landscape scales. *Frontiers in Ecology and the Environment*, **7**, 4–11.

Odum, E.P. (1971). *Fundamentals of Ecology*, 3rd edn. Saunders, New York.

O'Neill, R.V., DeAngelis, D.L., Waide, J.B. and Allen, T.F.H. (1986). *A Hierarchical Concept of Ecosystems*. Monographs in Population Biology, Vol. **23**. Princeton University Press, Princeton, New Jersey.

O'Riordan, T., Cooper, C., Jordan, A., et al. (1998). Institutional frameworks for political action. In: *Human Choice and Climate Change*, Vol. **1**. *The Societal Framework*, pp. 345–439. Battelle Press, Columbus, OH.

Pearce, D. and Moran, D. (1994). *The Economic Value of Biodiversity*. International Union for the Conservation of Nature and Natural Resources, Cambridge, UK.

Pearce, D.W. and Turner, R.K. (1990). *Economics of Natural Resources and the Environment*. BPCC Wheatsons, Exeter, UK.

Peterson, D. and Parker, V.T. (1998). *Ecological Scale: Theory and Application*, 1–34. Columbia University Press, New York, USA.

Post, D.M, Doyle, M.W, Sabo, J.L. and Finlay, J.C. (2007). The problem of boundaries in defining ecosystems: a potential landmine for uniting geomorphology and ecology. *Geomorphology*, **89**, 111–126.

Ricketts, T.H., Daily, G.C. and Ehrlich, P.R. (2004). Economic value of tropical forest to coffee production. *Proceedings of the National Academy of Sciences USA*, **101**, 12579–12582.

Savard, J.P.L., Clergeau, P. and Mennechez, G. (2000). Biodiversity concepts and urban ecosystems. *Landscape and Urban Planning*, **48**, 131–142.

Sayer, J. and Campbell, B. (2004). *The Science of Sustainable Development: Local Livelihoods and the Global Environment*. Cambridge University Press, Cambridge.

Smith, V. K. (1993). Nonmarket valuation of environment resources: an interpretative appraisal. *Land Economics*, **69**, 1–26.

Swift, M.J., Izac, A.M.N. and Noordwijk, M.V. (2004). Biodiversity and ecosystem services in agricultural landscapes-are we asking the right question? *Agriculture, Ecosystems and Environment*, **104**, 113–134.

Tacconi, L. (2000). *Biodiversity and Ecological Economics. Participation, Values, and Resource Management*. Earthscan, London.

Tobias, D. and Mendelsohn, R. (1991). Valuing ecotourism in a tropical rain-forest reserve. *Ambio*, **20**, 91–93.

Troy, A. and Wilson, M.A. (2006). Mapping ecosystem services: practical challenges and opportunities in linking GIS and value transfer. *Ecological Economics*, **60**, 435–449.

Turner, M.G., O'Neill, R., Gardner, R.H. and Milne, B.T. (1989). Effects of changing spatial scale on the analysis of landscape pattern. *Landscape Ecology*, **3**, 153–162.

Turner, W.R., Brandon, T. and Brooks, M. (2007). Global conservation of biodiversity and ecosystem services. *BioScience*, **57**, 868–873.

US Army Corps of Engineers, New York District (2004). *Hudson-Raritan Estuary. Liberty State Park Ecosystem Restoration Draft, Integrated Feasibility Report and Environmental Impact Statement*.

Van Jaarsveld, A.S., Biggs, R., Scholes, R., et al. (2005). Measuring conditions and trends in ecosystem services at multiple scales: the Southern African Millennium Ecosystem Assessment (SAfMA) experience. *Philosophical Transactions of the Royal Society Biological Sciences*, **360**, 425–441.

Van Koppen, C.S.A. (2000). Resource, arcadia, lifeworld. Nature concepts in environmental sociology. *Sociologia Ruralis*, **40**, 300–318.

Wilby, A. and Thomas, M.B. (2002). Natural enemy diversity and pest control: patterns of pest emergence with agricultural intensification. *Ecology Letters*, **5**, 353–360.

Xie, G., Lu, C, Leng, Y., Zheng, D. and Li, S. (2003). Ecosystem service assessment of Tibet high plateau. *Journal of Natural Resources*, **18**, 189–196.

Yue, T.X, Liu, J.Y. and Li, Z.Q. (2005). considerable effects of diversity indices and spatial scales on conclusions relating to ecological diversity. *Ecological Modeling*, **188**, 418–431.

Zhang, X. and Lu, X. (2010). Multiple criteria evaluation of ecosystem services for the Ruoergai Plateau Marshes in southwest China. *Ecological Economics*, **69**, 1463–1470.

8

Experimental Assessment of Ecosystem Services in Agriculture

Harpinder Sandhu,[1] John Porter[2] and Steve Wratten[3]

[1] School of the Environment, Flinders University, Australia
[2] Department of Plant and Environmental Science, University of Copenhagen, Denmark
[3] Bio-Protection Research Centre, Lincoln University, New Zealand

Abstract

Ecosystem services (ES) in agriculture are vital for the supply of food and fibre. However, the provision of some of the ES by these ecosystems has traditionally been considered to be at a low level. Earlier studies attributed very low values of ES to farmland world-wide per annum but the authors recognize that this was a severe underestimate because of the paucity of data available at the time. These assessments were based on published studies that used 'value transfer' techniques, supported by a few original calculations. In contrast to these studies, the current work proposes a framework and a 'bottom-up' approach to asses ES experimentally at field level. It elaborates on the conceptual framework of ES in agroecosystems providing field-scale assessments, citing examples from Denmark and New Zealand. This work demonstrates that there is a very wide range of ES provision, with organic arable cropping delivering many times the ES value of that provided by conventional farming. This study also provides scenarios for balancing production and ES in agroecosystems that can be explored to maintain and improve farm sustainability and achieve food security.

Introduction

Agriculture in the last century has evolved from self-sufficiency to surplus by growing more food per unit area. However, increased agricultural production has

Ecosystem Services in Agricultural and Urban Landscapes, First Edition. Edited by Steve Wratten, Harpinder Sandhu, Ross Cullen and Robert Costanza.
© 2013 John Wiley & Sons, Ltd. Published 2013 by John Wiley & Sons, Ltd.

resulted in global and local land use change (Vitousek et al., 1997; Goldewijk and Ramankutty, 2004; UNEP, 2005), ecosystem degradation and loss of ecosystem services (ES) (Heywood, 1995; Costanza et al., 1997; Daily, 1997; Krebs et al., 1999; Tilman et al., 2001). Despite unprecedented food production, more than one billion people are undernourished world-wide. Moreover, as human population adds another two billion by the middle of this century, there will be more stress on these agroecosystems to supply food. The United Nations has pledged to achieve Millennium Development Goals by 2015 that include eradication of hunger (UN, 2005). Agroecosystems cover 1.54 billion hectares world-wide and to meet the food demand of the growing population, an additional 0.4 billion hectares will be required. This has potential to increase agriculture's ecological footprint.

Current trends of agroecosystem degradation threaten to alter radically not only the capabilities to produce food and fibre but also the delivery of essential ES by these agroecosystems (Pretty, 2002). These nature's services or ES support life on earth through a wide range of processes and functions (Myers, 1996; Daily, 1997; Daily et al., 1997). ES provide major inputs to many sectors of the global economy and have been demonstrated to be of very high economic value (\$US33 trillion year^{-1}; Costanza et al., 1997). Yet because most of these services are not traded in economic markets, they carry no 'price tags' (no exchange value in spite of their high use value) that could alert society to changes in their supply or deterioration of underlying ecological systems that generate them. However, ES world-wide are being degraded more rapidly than ever before and this degradation poses serious threats to quality of life and therefore to modern economies. The Millennium Ecosystem Assessment (MEA, 2005) pointed to the very high rate of ES loss and the consequences for global stability if that rate continues. Thus the key challenge is to provide food security to a growing population and also to maintain and enhance the productivity of agroecosystems (UN, 1992). There is therefore currently an increasing interest in the utilization and enhancement of ES provided by agroecosystems.

In recent years, the concept of ES has gained wide acceptance within the international scientific community (Costanza et al., 1997; Daily, 1997; Tilman et al., 2002; Palmer et al., 2004; Robertson and Swinton, 2005; Sandhu et al., 2008, 2010a, 2012). It led to the adoption of the ES concept by the United Nations' sponsored Millennium Ecosystem Assessment (MEA) programme (www.millenniumassessement.org). Recently, to translate science into action world-wide, the United Nations has established the Intergovernmental Science-Policy Platform on Biodiversity and Ecosystem Services (IPBES, 2010).

In this chapter we discuss the framework of ES associated with agroecosystems and provide global examples of field-scale assessment of ES on farmland. Drawing on these assessments, we then build scenarios for production and ES associated with agriculture and concludes with recommendations for future research.

ES in agroecosystems

Agroecosystems being the largest managed ecosystems on Earth are often confronted by problems associated with ecosystem degradation. They contribute to the problem by consuming several ES and also offer solution as providers of ES. Much of the success of modern agriculture has been from provisioning ES such as food and fibre. However, the expansion of these marketable ES has resulted in the suppression of other valuable and essential ES such as climate regulation, water regulation, biodiversity, soil erosion protection etc. Maintaining these ES becomes vital in order to fulfil the food demand of the growing population, which will double by 2050. Therefore the need is to address the underestimation of ES in modified ecological systems such as farmland and explore methods of evaluating ES, as well as the ways in which ES in these systems can be maintained and enhanced.

ES associated with farming are classified into four groups (Table 8.1), based on the MEA (2005). Based on the ES literature and discussion with experts, several ES have been identified in agroecosystems (Cullen et al., 2004; MEA, 2005; EFTEC, 2005; Sandhu et al., 2007). Description of ES in the Millennium Assessment is based on natural ecosystems; therefore slightly different ES are identified in agroecosystems, which are discussed below.

Provisioning goods and services

These include food and services for human consumption, ranging from raw materials and fuel wood to the conservation of species and genetic material (de Groot et al., 2002; MEA, 2005). These goods and services are produced in agricultural landscapes by consuming some of the supporting and regulating services.

Supporting services

These are the services that are required to support the production of other ES. In this case they support food, fibre, feed and wood. Suppression of these support-ing ES can lead to their substitution with external inputs as is the case in substi-tuted agriculture where most of the supporting ES have been replaced by inputs or technology. Key supporting ES associated with agriculture are pollination, biological control, nutrient cycling and nitrogen fixation.

Regulating services

Ecosystems regulate essential ecological processes and life-support systems through biogeochemical cycles and other biospheric processes (Daily, 1997; Costanza et al., 1997). Hydrological flow in the plant–soil–atmosphere plays a critical role in arable farming. The hydrological cycle renews the earth's supply of water by distilling and distributing it (Gordon et al., 2005).

Cultural services

Cultural services contribute to the maintenance of human health and well-being by providing recreation, aesthetics and education (Costanza et al., 1997;

Table 8.1 Classification of ecosystem services (Costanza et al., 1997; de Groot et al., 2002; MEA, 2005; Sandhu et al., 2007).

Ecosystem services	Definition	Example
Regulating services		
1 Gas regulation	Regulation of atmospheric chemical composition	CO_2/O_2 balance, O_2 for UVB, SOx levels
2 Climate regulation	Regulation of global temperature, precipitation, and other biologically mediated climatic processes at global or local levels	Greenhouse gas regulation
3 Disturbance regulation	Capacitance, damping and integrity of ecosystem response to environmental fluctuations	Storm protection, flood control, drought recovery
4 Water regulation	Regulation of hydrological flow	Irrigation, milling transportation
5 Water supply	Storage and retention of water	Watersheds, reservoirs, aquifers
6 Erosion control and sediment retention	Retention of soil within an ecosystem	Erosion control, reduction of run-off
7 Waste treatment	Recovery of mobile nutrients and removal or breakdown of excess or xenic nutrients and compounds	Waste treatment, pollution control, detoxification
8 Refugia	Habitat for resident and transient production	Nurseries, habitat for migratory species, regional habitats for locally harvested species
Provisioning services		
9 Food production	That portion of gross primary production extractable as food	Production of fish, crops, nuts, fruits
10 Raw material	That portion of gross primary production extractable as raw material	Production of lumber, fuel or fodder
11 Genetic resources	Sources of unique biological materials and products	Products for materials science, resistance to plant pathogens and crop pests

(continued)

Table 8.1   (cont'd)

Ecosystem services	Definition	Example
12 Ornamental resources	For display purpose	Horticultural products, flowers, etc.
13 Medicinal resources	Source of medicinal compounds	Products used in medicines
Cultural services		
14 Aesthetic information	Associated landscapes	Landscaping of farmland
15 Recreation	Providing opportunities for recreational activities	Ecotourism, sport fishing, outdoor activities
16 Cultural and artistic information	Providing opportunities for non-commercial uses	Aesthetic, artistic, education spiritual, and/or scientific values
17 Spiritual and historic information	Source of historic and spiritual value	Associated history of farmsteads
18 Science and education information	Source of education and training	Research and development
Supporting services		
19 Pollination	Movement of floral gametes	Reproduction of plant populations
20 Biological control	Trophic–dynamic regulations of population	Reduction of herbivory by top predators, control of prey species
21 Carbon accumulation	Carbon sequestration by vegetation and soil	Regulation of chemical composition
22 Mineralization of plant nutrients	Storage, internal cycling, processing and acquisition of nutrients	Nitrogen fixation
23 Soil formation (maintenance of soil health)	Soil formation processes (turning over of soil by earthworms	Structure maintenance
24 Nitrogen fixation	Storage and cycling	Legumes fixing N
25 Services provided by shelterbelts	Protection against wind erosion	Windbreaks

de Groot et al., 2002; MEA, 2005). Agriculture provides these services as some farmers conserve field-boundary vegetation or enhance landscapes by planting hedgerows, shelterbelts or native trees. Some farms provide accommodation and recreational activities for family members as well as for national and international visitors. Participation of farms in research and education enhances this cultural service (Warner, 2006). Agricultural landscapes also have cultural heritage value.

Field-scale assessment of ES

Recent work has estimated the value of global ecosystem goods and services (Costanza et al., 1997; de Groot et al., 2002; MEA, 2005), generating increased awareness of their classification, description, economic evaluation and enhancement (Gurr et al., 2004). This valuation is heavily weighted towards natural ecosystems biomes, such as boreal forests, coral reefs, mangroves etc. and, in fact, attributed no dollar value to highly modified or 'engineered' ecosystems such as farmland, forestry and cities rather than 'engineered' ones, which are actively modified by humans (Balmford et al., 2002).

These assessments were based on published studies and used 'value transfer' techniques, supported by a few original calculations. These studies provoked meaningful debate about appropriate ways to value ES (Toman, 1998; Turner et al., 1998; Farber et al., 2002). Some contributors to the debate have argued that attempts to provide estimates of the value of global ES are misguided as there is no potential purchaser of the total ES (Dasgupta et al., 2000). In contrast, some authors argue there is merit in estimating the incremental changes of values in ES at specific sites and locations (Turner et al., 1998).

Therefore, in contrast to the above methods for the valuation of ES, the following section describes and discusses quantification of the economic value of ES in highly modified and productive farming landscapes ('engineered systems') in Denmark and New Zealand using a 'bottom-up' approach. It demonstrates the value in the arable sector for the maintenance of profit and sustainable practices by addressing both conventional as well as organic systems. Field assessment of ES in a Danish combined food and energy system (CFE; Porter et al., 2009) and also in New Zealand arable farmland (Sandhu et al., 2008; 2010a) is discussed here. ES were identified and measured by field-scale processes and translated into monetary terms by using willingness-to-pay, value-transfer and avoided cost estimates. Willingness-to-pay is the maximum amount an individual is willing to pay to achieve a specific goods or service. Value transfer is an economic methodology which obtains an estimate for the economic value of non-market goods or services through the analysis of a single study, or group of studies, that have been previously carried out to value similar goods or services. The 'transfer' itself refers to the application of economic values and other information from the original work to a new study or synthesis (e.g. Costanza et al., 1997). Avoided cost estimates allow society to avoid costs that would have been incurred in the absence of those ecosystem services. For example, biological control provided by natural enemies of pests avoids the variable costs (labour, pesticides, diesel etc.) of using pesticides. We can use this to estimate a value for the biological control, soil services etc.

Field-scale assessment of ES in agriculture can help in redesigning agricultural landscapes using new ecotechnologies based on novel and sound ecological knowledge to enhance ES. Ecotechnologies, such as enhancing mineralization

(a)

(b)

Fig. 8.1   (a) CFE with biofuel belts and crops. (b) New Zealand arable farm with shelterbelts.

of plant nutrients by managing stubble plant residue after harvest and incorporating flowering plants to provide nectar source to parasitic wasps to enhance biological control of insect pests, are some of the examples. This helps to ensure long-term sustainability of farms in the face of very rapid human population growth.

The combined food and energy system

The combined food and energy (CFE) system study site is at the experimental farm of the University of Copenhagen, Denmark. It consists of 10.1 ha of arable food (barley and wheat) and a pasture fodder crop (clover-grass), and ca. 1 ha of biofuels, which consists of four belts of fast-growing trees (willows, alder and hazel) (Fig. 8.1a). This system is a net energy producer, with the system producing more energy in the form of renewable biomass than consumed in the planting, growing and harvesting of the food and fodder (Porter et al., 2009).

Coincidentally with the issue of ES from agroecosystems, there is a developing interest in using agricultural land for the production of biofuels (Tilman et al., 2006) such that their production is as sustainable as possible. Such a requirement invites the design of new systems of primary production that ensure a positive net carbon sequestration, are species diverse, have low inputs and provide a suite of ES. An experimental example of such a system is a CFE producing agroecosystem that meets the above requirements for sustainability by using non-food hedgerows as sources of biodiversity and biofuel. Previous work (Porter et al., 2009) has identified, quantified and valued ES from this production system and refers to this concept as combined food, energy and ecosystem services (CFEES). The bioenergy component in the CFE system is represented by belts of fast-growing trees (willows, alder and hazel) that are planted orthogonally to fields containing cereal and pasture crops and the system is managed organically, meaning that biocides and inorganic nitrogen are not used.

The Millennium Assessment (MEA, 2005) reported loss of ES world-wide and more recent reports (FAO, 2007; Steinfield et al., 2006) advocate designing production systems that can contribute towards global ES. The CFE agroecosystem provides a novel way of producing food, fodder and energy in the form of renewable biomass and ES. Field-based estimates of individual ES identified in this system provides conservative estimates of the economic value (Fig. 8.2). The value of supporting services (biological control of pests, N regulation (fixation and mineralization), soil formation, carbon accumulation and pollination) and regulating services (hydrological flow) is based on avoided cost estimates. Provisioning services included food and fodder production and biomass production. Their value is based on farm gate prices of produce. Aesthetics ES, identified in this study as cultural service, was assessed using value-transfer as no other estimate was available. Agri-environment schemes implemented in various EU countries do not effectively yield outcomes to balance farming activities and environmental outcomes (Foley et al., 2005). The CFE system offers scope to maintain this balance.

New Zealand arable farmland

The role of land-management practices in the maintenance and enhancement of ES in agricultural land is investigated by quantifying the economic value of ES at the field level based on an experimental approach. The study sites included 29 arable fields, distributed over the Canterbury Plains in New Zealand and comprised 14 organic and 15 conventional fields (Sandhu et al., 2008, 2010a; Fig. 8.1b). First, the role of land-management practices in the maintenance and enhancement of ES in agricultural land was investigated by quantifying the economic value of ES at the field level under organic and conventional arable

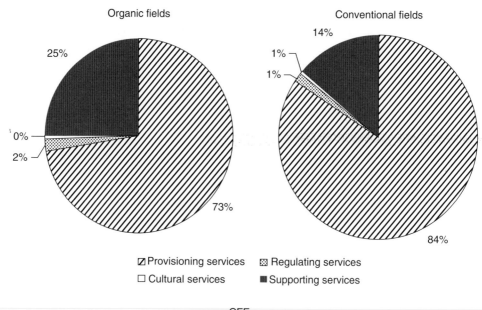

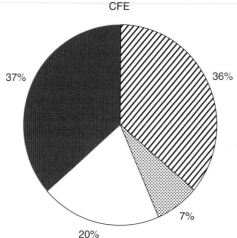

Fig. 8.2 Summary of mean economic value of ecosystem services in organic and conventional fields in New Zealand arable land and Danish combined food and energy (CFE) system. Percentage of each group of ES (provisioning, supporting, regulating and cultural services) out of the total economic value in three different systems is shown here.

systems. Total economic value of ES in organic fields ranged from $US1610 to 19420 ha^{-1} year^{-1} and that of conventional fields from $US1270 to 14570 ha^{-1} year^{-1}. The non-market value of ES in organic fields ranged from $US460 to 5240 ha^{-1} year^{-1}. The range of non-market values of ES in conventional fields was $US50 to 1240 ha^{-1} year^{-1}. There were significant differences between organic and conventional fields for the economic values of some ES. Next, this economic information was used to extrapolate and to calculate the total and non-market value of ES in Canterbury arable land. The total annual economic and non-market values of ES for the conventional arable area in Canterbury (125 000 ha) were $US332 million and $US71 million, respectively. If half the arable area under conventional farming shifted to organic practices, the total economic value of ES would be $US192 million and $US166 million annually for organic and conventional arable area, respectively. In this case, the non-market value of ES for the organic area was $US65 million and that of the conventional area was $US35 million annually. This study demonstrated that arable farming provides a range of ES which can be measured using field experiments based on ecological principles by incorporating a 'bottom-up' approach.

The benefits of ES in 'engineered' ecosystems are substantial as demonstrated by their economic value in arable land in Canterbury, New Zealand (Fig. 8.2). The ecological and economic value of some of the ES can be maintained and enhanced on arable farmland by adopting sustainable practices such as organic farming (Sandhu et al., 2010b). This study makes clear that arable farmland provides a range of ES which can be measured using field experiments based on ecological principles by incorporating a 'bottom-up' approach.

Scenarios of production and ES in agroecosystems

The manner in which global farming may affect ES on farmland in the future depends partly on the range of plausible scenarios of agriculture. An urgent scientific challenge is to examine the impacts of global farming on ES and how to ameliorate them. This depends partly on the range of plausible scenarios for future agriculture. As with other visioning exercises (MEA, 2005), there are many possible scenarios but for agriculture we propose a model of eight scenarios based on the experimental assessment of ES (Fig. 8.3). This explores plausible futures for ES and production in agroecosystems. Agriculture being the single-largest human driven ecological activity on earth has the potential to degrade ecosystems or to enhance them. Agriculture has been very successful in achieving outputs or providing ES for which markets exist but often at the expense of other essential and vital ES. Scenarios obtained above are explained under the following five systems.

The ethnocentric systems

This system advocates that cultural and social values are foremost and overlooks the economic growth that is required to support the growing population. S1 scenario falls under this system (Fig. 8.3). Here the output, provision of ES and management costs associated with it are low and can be considered as poor systems.

The technocentric systems

The technocentric systems are based on the assumptions that there are limitless natural resources and can be exploited for unhindered production. It does not recognize limited capacity of the planet to supply these resources and the

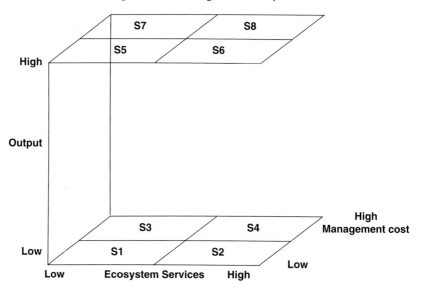

Fig. 8.3   Eight scenarios (S1–S8) for ES production and management cost in relation to output in agroecosystems. See text for details.

Scenario	Output	Ecosystem services	Management costs	Agroecosystems	Category
S1	Low	Low	Low	Poor systems	Ethnocentric
S2	Low	High	Low	Ideal for environmental outcomes	Ecocentric
S3	Low	Low	High	Degraded systems	Technocentric
S4	Low	High	High	Multifunctional systems	Sustaincentric
S5	High	Low	Low	Current monocultures	Technocentric
S6	High	High	Low	Optimal systems	Ecotechnocentric
S7	High	Low	High	Changing ecosystems	Ecocentric
S8	High	High	High	Unsustainable	Technocentric

interdependence between human capital and natural capital. Modern agriculture since the beginning of industrial revolution is based on this and has resulted in immense production of food and fibre but it is surrounded by ecosystems degradation. Scenarios S3, S5 and S8 fall under this category (Fig. 8.3). In S3, output and ES are low whereas management costs are high and are termed as degraded systems. In S5, which represents current monocultures, output is projected to be high but ES provision and management costs are low. Scenario S8 is associated with high output, ES provision and management costs and is unsustainable.

The ecocentric systems

This view advocates that humans are part of the web of life and earth is governed by self-regulating mechanisms (Devall and Sessions, 1985). This system argues for having perfect relationship with nature, which has an intrinsic value independent of human values (Leopold, 1949). However, it is not a practical paradigm. S2 scenario is ideal for environmental outcomes with low output and management costs but higher provisions of ES, and falls under this category. Scenario S7, which represents changing ecosystems, with high output and management costs and low ES provision, also falls under this category.

The ecotechnocentric systems

This system takes into consideration both ecocentric and technocentric systems by incorporating economic growth as well as including natural resources management. S6 describes optimal systems with high output and ES provision and low management costs, and falls under this category.

The sustaincentric systems

The sustaincentric system is based on the belief of have a holistic and balanced system. Although it addresses the necessity of economic development, it also incorporate the consequences of depleting natural resources and overflowing sinks which have resulted from unabated economic growth. The sustaincentric paradigm incorporates the fact that ecological wealth underpins economic wealth. S4 scenario is best to address this view in agroecosystems. It is defined as low output and higher ES provision with higher management costs and can be termed multifunctional systems.

Conclusions

Farmland provides a range of ES which can be measured using field experiments based on ecological principles by incorporating a 'bottom-up' approach, as discussed in the sections above. Evaluation of ES provides information for policy and decision makers to consider the financial contribution of different farming practices towards the sustainability of agriculture. The 'substitution' agriculture currently in practice has resulted in degradation of some ES to such an extent that they have no economic value on these 'engineered' or designed landscapes. The challenge of reversing the degradation of these ES can be partially met by practising 'ecological engineering' under some alternative form of land-management practices.

The economic values reveal significant changes in ES in monetary terms and help ensure that arable farming is a contributor to improved social well-being as well as increased food production. Perspectives for further work may include research to improve understanding of the basis of ecological processes and mechanisms to understand the trade-offs and synergies provided by different land-management practices (conventional, organic or conservation agriculture). Further research can evaluate and make recommendations for the 'best management practices' to enhance ES and reduce net externalities from agricultural system. Future research can also focus on the operation and behaviour of combined multifunctional cropping systems for food and fibre on local biodiversity and the system's economic and energy balance in terms of its fossil and renewable energy use and production.

References

Balmford, A., Bruner, A., Cooper, P., et al. (2002). Economic reasons for saving wild nature. *Science*, **297**, 950–953.

Costanza, R., d'Arge, R., De Groot, R., et al. (1997). The value of the world's ecosystem services and natural capital. *Nature*, **387**, 253–260.

Cullen, R., Takatsuka, Y., Wilson, M. and Wratten, S. (2004). *Ecosystem Services on New Zealand Arable Farms*, pp. 84–91. Agribusiness and Economics Research Unit, Lincoln University, Discussion Paper 151.

Daily, G.C. (1997). *Nature's Services: Societal Dependence on Natural Ecosystems*, pp. 1–10. Island Press, Washington, DC.

Daily, G.C., Alexander, S., Ehrlich, P.R., et al. (1997). Ecosystem services: benefits supplied to human societies by natural ecosystems. *Issues in Ecology*, **2**, 18.

Dasgupta, P., Levin, S. and Lubchenco, J. (2000). Economic pathways to ecological sustainability. *BioScience*, **50**, 339–345.

de Groot, R.S., Wilson, M. and Boumans, R.M.J. (2002). A typology for the classification, description and valuation of ecosystem functions, goods and services. *Ecological Economics*, **41**, 393–408.

Devall, B. and Sessions, G. (1985). *Deep Ecology*. Gibbs Smith, Salt Lake City.

EFTEC (2005). *The Economic, Social and Ecological Value of Ecosystem Services*. Available at: http://www.jncc.gov.uk/pdf/BRAS_SE_Newcomeetal-TheEconomic,SocialandEcologicalValueofEcosystemServices(EftecReport).pdf (accessed August 2012).

Farber, S.C., Constanza, R. and Wilson, M.A. (2002). Economic and ecological concepts for valuing ecosystem services. *Ecological Economics*, **41**, 375–392.

FAO (2007). *The State of Food and Agriculture: Paying Farmers for Environmental Services*, Series No. 38. FAO Agriculture, Rome.

Foley, J.A., deFries, R., Asner, G.P., et al. (2005). Global consequences of land use. *Science*, **309**, 570–573.

Goldewijk, K.K. and Ramankutty, N. (2004). Land cover change over the last three centuries due to human activities: the availability of new global data sets. *GeoJournal*, **61**, 335–344.

Gordon, L.J., Steffen, W., Jonsson, B.F., Folke, C., Falkenmark, M. and Johannessen, A. (2005). Human modification of global water vapour flows from the land surface. *Proceedings of the National Academy of Sciences USA*, **102**, 7612–7617.

Gurr, G.M., Wratten, S.D. and Altieri, M.A. (eds) (2004). *Ecological Engineering for Pest Management: Advances in Habitat Manipulation for Arthropods*. CSIRO, Victoria.

Heywood, V.H. (ed.) (1995). *United Nations Environment Program, Global Biodiversity Assessment*. Cambridge University Press, Cambridge.

IPBES (2010). *Intergovernmental Science-Policy Platform on Biodiversity and Ecosystem Services*. UNEP. Available at: http://ipbes.net/ (accessed August 20112).

Krebs, J.R., Wilson, J.D., Bradbury, R.B. and Siriwardena, G.M. (1999). The second silent spring? *Nature*, **400**, 611–612.

Leopold, A. (1949). *A Sand Country Almanac*. Oxford University Press, New York.

MEA (2005). *Millennium Ecosystem Assessment Synthesis Report*. Island Press, Washington, DC.

Myers, N. (1996). Environmental services of biodiversity. *Proceedings of the National Academy of Sciences USA*, **93**, 2764–2769.

Palmer, M., Bernhardt, E., Chornesky, E., et al. (2004). Ecology for a crowded planet. *Science*, **304**, 1251–1252.

Porter, J., Costanza, R., Sigsgaard, L., Sandhu, H. and Wratten, S. (2009). The value of producing food, energy and ecosystem services within an agro-ecosystem. *Ambio*, **38**, 186–193.

Pretty, J. (2002). *Agri-Culture: Reconnecting People, Land and Nature*. Earthscan, London.

Robertson, G.P. and Swinton, S.M. (2005). Reconciling agricultural productivity and environmental integrity: a grand challenge for agriculture. *Frontiers in Ecology and the Environment*, **3**, 38–46.

Sandhu, H.S., Crossman, N.D. and Smith, F.P. (2012). Ecosystem services and Australianagricultural enterprises. *Ecological Economics*, **74**, 19–26.

Sandhu, H.S., Wratten, S.D. and Cullen, R. (2007). From poachers to gamekeepers: perceptions of farmers towards ecosystem services on arable farmland. *International Journal of Agricultural Sustainability*, **5**, 39–50.

Sandhu, H.S., Wratten, S.D., Cullen, R. and Case, B. (2008). The future of farming: the value of ecosystem services in conventional and organic arable land. An experimental approach. *Ecological Economics*, **64**, 835–848.

Sandhu, H.S., Wratten, S.D. and Cullen, R. (2010a). The role of supporting ecosystem services in arable farmland. *Ecological Complexity*, **7**, 302–310.

Sandhu, H.S., Wratten, S.D. and Cullen, R. (2010b). Organic agriculture and ecosystem services. *Environmental Science and Policy*, **13**, 1–7.

Steinfield, H., Gerber, P., Wassenaar, T., Castel, V., Rosales, M. and de Haan, C. (2006). *Livestock's Long Shadow: Environmental Issues and Options*. Livestock, Environment and Development Initiative. FAO, Rome.

Tilman, D., Cassman, G., Matson, P.A., Naylor, R. and Polasky, S. (2002). Agricultural sustainability and intensive production practices. *Nature*, **418**, 671–677.

Tilman, D., Fargione, J., Wolff, B., D'Antonio, C., Dobson, A., Howarth, R., et al. (2001). Forecasting agriculturally driven global environmental change. *Science*, **292**, 281–284.

Tilman, D., Hill, J. and Lehman, C. (2006). Carbon-negative biofuels from low-input high- diversity grassland biomass. *Science*, **314**, 1598–1600.

Toman, M. (1998). Why not to calculate the value of the world's ecosystem services and natural capital. *Ecological Economics*, **25**, 57–60.

Turner, R.K., Adger, W.N. and Brouwer, R. (1998). Ecosystem services value, research needs, and policy relevancy: a commentary. *Ecological Economics*, **25**, 61–66.

UN (1992). *Promoting Sustainable Agriculture and Rural Development*. United Nations Conference on Environment and Development. Rio de Janeiro, Brazil, 3 to 14 June. Agenda 21, 14.1–14.104. Available at: http://www.un.org/esa/sustdev/agenda21.htm (accessed August 2012).

UN (2005). *The Millennium Development Goals Report*. United Nations, New York.

UNEP (2005). *One Planet, Many People: Atlas of Our Changing Environment*. United Nations, New York.

Vitousek, P.M., Mooney, H.A., Lubchenco, J. and Melillo, J.M. (1997). Human domination of earth's ecosystems. *Science*, **277**, 494–499.

Warner, K.D. (2006). Extending agroecology: grower participation in partnerships is key to social learning. *Renewable Agriculture and Food Systems*, **21**, 84–94.

Part D

Designing Ecological Systems to Deliver Ecosystem Services

9

Towards Multifunctional Agricultural Landscapes for the Upper Midwest Region of the USA

Nicholas Jordan[1] and Keith Douglass Warner[2]

[1]Agronomy and Plant Genetics Department, University of Minnesota, St Paul, MN, USA
[2]Center for Science, Technology and Society, Santa Clara University, Santa Clara, CA, USA

Abstract

New strategies of agricultural research and development are needed to increase agricultural productivity, profitability and resilience to sustainability challenges from variable rainfall, energy costs and other unpredictable events. Development of a more multifunctional agriculture can reconcile key interests of environmental, economic and agricultural sectors of society and thus provides such a strategy. We outline a heuristic version of this strategy, involving coordination of scientific and social action and integration across multiple scales, sectors and systems. We illustrate implementation of the strategy via a case study.

Introduction

To address the interlocking challenges of climate change, sustainable management of agriculture and bioresources, human society will have to devise an integrated strategy to help multiple social actors and institutions make decisions guided by sustainability goals. In this chapter we propose such a strategy, in the context of sustainable agriculture. Our intention is to address the interlocking challenges by responding to a rapid increase in the social importance of agriculture. Rising interest in biofuels is the most visible indicator of this trend, but it has many other manifestations. In addition to major increases in global food production, society is now demanding that agriculture

Ecosystem Services in Agricultural and Urban Landscapes, First Edition. Edited by Steve Wratten, Harpinder Sandhu, Ross Cullen and Robert Costanza.
© 2013 John Wiley & Sons, Ltd. Published 2013 by John Wiley & Sons, Ltd.

produce a wide range of other goods, services and amenities (Meyer et al., 2008). In addition to biofuels, these include various bioindustrial products and marketable environmental services produced by agriculture, such as carbon storage, biodiversity conservation and aquifer recharge inputs (Boody et al., 2005; Eaglesham, 2006; Jordan et al., 2007). In essence, the challenge to agriculture is to increase production of food, biofuels and bioindustrial feedstocks, while maintaining the integrity of essential life-support functions of the biosphere. This difficult project must make progress in the face of global environmental change, which may include rapid climate change. Taken together, these intertwined issues of production, conservation and adaptation most assuredly constitute one of the 'grand challenges' facing humanity.

To meet this challenge, it will be necessary to substantially redesign agricultural production systems and their interface with food, water and energy systems. In response, multifunctional agriculture (MFA) is emerging. In essence, MFA is a project of 'sustainable land architecture,' which seeks complex land-use/land-cover systems that can meet multiple human needs from diverse ecosystems while sustaining these systems over multiple generations (Turner et al., 2007). MFA is defined by joint production of both agricultural commodities and a range of ecological services. These services include beneficial effects on pest and nutrient management, water quality and quantity, biodiversity and amenity values.

Despite its promise, adoption of MFA in the USA is not yet extensive; in our view, adoption has been impeded by sociopolitical, economic and ecologic factors that are interrelated and mutually reinforcing. A comprehensive approach to surmounting such barriers is therefore essential. In this essay, we propose an integrative and heuristic strategy – a 'theory of change' – that aims to increase the multifunctionality of US agriculture by pursuing change at three distinct levels of integration. The strategy capitalizes on certain distinctive attributes of multifunctional agroecosystems.

Multifunctional agroecosystems

More sophisticated agroecosystem designs can increase the multifunctionality of agricultural landscapes. These designs feature diverse perennial crops, grown on environmentally sensitive sites such as riparian areas. As well, non-commodity cover crops are grown after annual field crops. Multifunctionality arises as an emergent property of sustainable land architecture, that is the spatial and temporal pattern of perennial, annual and cover crops across landscapes and the resultant ecological processes. For illustrative purposes, we discuss potential applications of MFA systems to the Upper Midwest region of the USA, a globally important region comprising ca. 135 million acres of cropland.

In MFA systems for this region, production of annual crops will be complemented by strategic use of perennial-based agroecosystems. Perennial system well suited for this region include woody and herbaceous perennial polycultures, agroforestry systems and managed wetlands (Hey et al., 2005; Jorgensen et al., 2005; Tilman et al., 2006). Well-designed MFA landscapes that include both annual and perennial production systems can produce agricultural commodities abundantly and profitably, while also producing non-market public goods and services effectively. Examples of the latter include: (a) soil and nitrogen loss rates from perennial crops are less than 5% of those in annual crops; (b) perennial cropping systems have greater capacity to sequester greenhouse gases than annual-based systems; (c) in certain scenarios, some perennial crops appear more resilient to climate change than annuals, for example increases of 3 to 8°C are predicted to increase North American yields of the perennial crop switchgrass (*Panicum virgatum*), whereas declines are expected for annual crops; and (d) among species of concern for conservation, 48% increased in abundance when on-farm perennial land cover was increased in European Union 'agroenvironmental' incentive programmes (Gantzer et al., 1990; Brown et al., 2000; Robertson et al., 2000; Kleijn et al., 2006).

Thus, these new landscape designs address challenges that have resulted from the agricultural intensification of Upper Midwest landscapes during the past 50 years. Intensification has resulted from a range of land-use/ land-cover change. Mixed farming systems containing areas of low-intensity land use have been replaced by systems focusing on high-intensity specialized land use. Field sizes have increased, crop growth is limited to a 4- to 6-month period, and these landscapes are otherwise largely unvegetated. Via intensification, spatial and temporal patterns of land cover and land use have been greatly simplified. Effects include removal of perennial vegetation from pastures and field edge habitats (e.g. fence rows, riparian buffers). The hydrology of agricultural landscapes has been modified on a regional scale through agricultural drainage systems that lower water tables and which may exacerbate drought and flood risks and compromise water quality by discharges of nutrients, sediment and flood waters to surface and ground waters (Goolsby et al., 1999; Donner and Kucharik, 2008).

Consequently, the value of these landscapes for biodiversity conservation has been reduced. Fish and game species and other elements of biodiversity have been affected by habitat loss and fragmentation, pesticide pollution and invasive exotic species. Also, ecological services of high value to agricultural production have likely been reduced; these include soil protection and the uptake of water and nutrients, regulation of pest population levels and pollination. Current farming systems may also face significant challenges to their economic viability through increased input costs, new pests and diseases, soil degradation, groundwater depletion and increasing risk from climate extremes. More complex and multifunctional landscape designs have potential to provide cost-effective solutions to these conservation and production problems (Hanson et al., 2007; Zhang et al., 2007).

Re-designed agricultural landscapes for the Upper Midwest

In the Upper Midwest, there is a crucial need to develop agricultural landscapes that are capable of supporting intensified crop production while also producing ecosystem services at high levels and conserving soil, water and biodiversity resources. A substantial base of evidence suggests that these goals could be accomplished by creating functionally diverse landscapes composed of annual and perennial cropping systems, strategically placed to provide continuous living cover across a high proportion of the landscape. These landscapes will be configured to buffer conservation functions from production functions as much as possible, and grassland, wetland and other perennial-based agroecosystem areas will be spatially configured for high conservation value. Of particular interest is the design, development and management of MFA landscapes at scales on the order of 25 km². This spatial scale is critically important because landscapes re-designed and managed at such scales may offer major opportunities, through economies of scale and scope, to increase outputs and lower production costs for both agricultural commodities and ecological services.

We propose that re-designed MFA landscapes for the Upper Midwest will have the following attributes:

1   Agricultural production is practised on as much of the landscape as possible.
2   Substantial areas are allocated to annual agriculture with continuous living cover to improve soil quality, store C and support intensified production.
3   Perennial production systems (e.g. managed grassland and wetland systems) are located on sites poorly suited to annual agriculture.
4   A landscape pattern is designed and implemented that creates large-scale upland networks of perennial species (woody and herbaceous) that regulate and ameliorate climate, conserve wildlife and biodiversity, store carbon, and reduce methane and nitrous oxide.
5   A landscape pattern is designed and implement to create large-scale lowland networks of perennial vegetation surrounding and interconnecting waterways and wetlands, to regulate water quality and flow, and support groundwater recharge.

Moving forward on design and implementation of multifunctional landscapes for the Upper Midwest

Multifunctional landscapes are also favoured by a number of other trends in food and energy, providing new economic opportunities for farmers and landowners. The rapid development of bioenergy production could provide a market force strong enough to establish perennial 'energy crops' across substantial acreages in the Midwest landscape. Demand for grass-fed meat and dairy has increased dramatically, and markets are emerging for high-value products from new perennial and woody crops, such as perennial flax and hazelnuts. MFA systems may enable relatively small farm units to respond to these new economic opportunities, which may, in turn, provide the capital and population base necessary for healthy, productive and dynamic rural communities. An integrative assessment of the potential economic, social and environmental performance of MFA designs in the Upper Midwest is provided by a modelling study. Model simulations projected that major environmental and socioeconomic benefits could be attained from MFA designs without an increase in public subsidy costs (Flora, 2001; Boody et al., 2005; Argent et al., 2007).

To capitalize on these opportunities and address current problems, we suggest that systemic change is needed in multiple social and economic sectors. To date, environmental, economic and social problems arising from current Midwest agriculture have been addressed in a fragmentary way, mainly through policies that subsidize the retirement of environmentally sensitive sites from production, and by promotion of 'best management practices' that optimize production methods for annual crops. These approaches have made some gains, but subsidies are limited by their high cost per unit area, while optimized annual crop production cannot adequately address environmental problems. For example, despite optimal or suboptimal fertilizer application rates, nutrient losses from annual crop production often exceed acceptable levels (Magner et al., 2004).

Rather, comprehensive and coordinated change is needed in social and biophysical dimensions of Midwest agriculture. In particular, economic incentives must be reconfigured to achieve multiple social goals, and new modes of perception, knowledge production and decision making are needed. These innovations are needed to develop the necessary policies and markets to stimulate a diversified flow of goods and services from multifunctional agricultural landscapes. Below we outline a systemic strategy for increasing multifunctionality in Midwest agriculture.

Theory of change: a social–ecological system model for increasing multifunctionality of agricultural landscapes

Our model encompasses three distinct dynamic processes, operating at different space/ time scales. We first describe the central 'system' level, which addresses 'enterprise development', that is the development of new economic opportunities, and related systems of management and policy, for farmers of multifunctional agroecosystems. Next, we portray a pivotal subsystem of the enterprise development model, 'agroecological partnerships', that produce knowledge needed for multifunctionality in working agroecosystems. Finally, we describe how social values shape the 'supersystem' of public opinion and policy, and how this can be configured to reward increased multifunctionality in agriculture. Our approach reflects recent theoretical advances in the integration of natural and social science with sustainable development processes, as well as the integration of development with social and biophysical change processes at various scales (Gunderson and Holling, 2002; Turner et al., 2007; Holtz et al., 2008).

Focal level: enterprise development via 'virtuous circles'

We draw upon an emerging theory of endogenous (i.e. 'bottom-up') rural development, termed 'virtuous circles'. Virtuous circles of rural development are positive feedback loops that integrate and enhance rural resources, including assets that are natural, human, social, cultural, political and financial in nature. In principle, the virtuous circle process generates synergy between natural resources situated in MFA systems and other resources, resulting in their joint increase (Selman and Knight, 2006).

The operation of a virtuous circle process (Fig 9.1) is based on effective joint production of traditional agricultural commodities and other valuable goods, services and amenities. To the extent that joint production occurs, there will be the opportunity to capture value from both of these streams of production, and a variety of sectors will have incentive to do so. For example, in Minnesota, recent analyses indicate that restoration of ca. 2 million acres of high-quality duck habitat will be needed to restore duck populations and address widespread concern with recent population declines. Current approaches cannot achieve this goal, so state and non-governmental organization conservation agencies are increasingly focused on MFA designs that improve the quality and quantity of duck habitat in agricultural landscapes. This approach appears to offer substantial cost savings to conservation agencies.

We posit that in situations where such opportunities are recognized and transaction costs are low, multiple interest groups (e.g. wildlife groups, farmer organizations and renewable energy interests) will be willing to co-operate in support of multifunctional agroecosystems that can provide amenities, goods and services to these groups on attractive terms. When cooperation is effective, new forms of support for multifunctional agroecosystems appear. Such support may result in new markets for commodities, amenities, goods and services from multifunctional agroecosystems or in supportive non-market mechanisms such as 'farmland protection programmes' and other policy measures at local or regional levels. Co-operation may also increase support for these agroecosystems provided by a wide range of sectors (e.g. banking, regulatory, technical, commercial, educational). In principle, these flows of revenue and resources to landowners and managers of multifunctional agroecosystems serve to close the positive feedback loops that drive the virtuous circle process, thereby increasing the adoption and extent of multifunctional agroecosystems.

The virtuous circle model highlights several aspects of development of MFA that are underappreciated, in our experience. Most importantly, enterprise development based on MFA entails change in coupled human–environment systems understood as webs of interdependent causal factors. These are social, economic and biophysical in nature and operate over a wide range of space/time scales. Such causal webs are place and problem specific and so also are the challenges of enterprise development. Given this level of complexity, we conclude that enterprise development will require creation of new place-based institutions – 'adaptive co-management systems' (ACM) (Armitage et al., 2007). ACM systems aim to create shared understanding of social and biophysical causal factors among multiple stakeholders, and take better advantage of subsystem interdependencies and the capacity for concerted and coordinated action among stakeholders to address complex challenges such as enterprise development, with its interacting social and biophysical dimensions (Pahl-Wostl and Hare, 2004). In our view, ACM is a critical part of the implementation of our theory of change. We posit that the virtuous circle process can drive enterprise development only if there is some agency that can insure that a rapid positive feedback process can occur; in our view, ACM provides that agency.

However, social and biophysical factors oppose the formation and effective operation of ACM systems. To date, ACM has evolved in rudimentary form in a

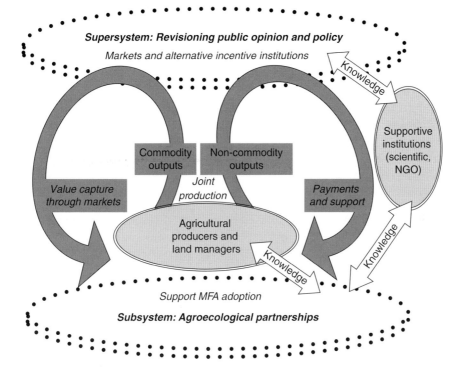

Fig 9.1 Multifunctional agriculture (MFA) enterprise development via the virtuous circle model. The model features central feedback loops connecting agricultural producers to markets and incentive institutions (e.g. brokers for ecological services produced by MFA) via joint production of agricultural commodities and non-commodities outputs, such as ecological services. Stakeholders, markets and institutions provide a range of forms of payment and support for MFA. Enterprise development interacts with the 'supersystem' of public opinion and policy, as stakeholders attempt to change this super-system to increase support for MFA. Agroecological partnerships form a subsystem that supports enterprise development by increasing multifunctionality.

range of situations, but its spatial and temporal 'area of influence' has been limited in most cases (Plummer and Armitage, 2007). These limitations particularly result from major challenges to ACM. First, there is the difficulty of necessary but unfamiliar forms of comprehension, cognition and learning. ACM requires rich systemic understanding of biophysical factors, social factors and their interplay, which depends on effective integration across a wide range of existing knowledge. Equally, it depends on creation of new knowledge through the effective interplay of multiple rationalities and knowledge systems, held by multiple stakeholders. New methods are urgently needed that better support such multiple and interactive processes of comprehension, cognition and learning in ACM.

As well, the power and resilience of established management and land-use systems creates a powerful barrier. ACM must be able to establish itself in the face of 'stabilizing factors' – both biophysical and social – that reinforce established management systems that do not support ACM. Such 'stabilizers' include path dependencies of existing land-use management strategies, the opposition of dominant political actors and cumulative effects that are distant in time and space from land-use decisions that generate them. At present, there is little theoretical understanding of how ACM might interact with biophysical and social factors to become established and broadly improve the sustainability of land use (Armitage et al., 2007).

Second, since substantial costs and risks are of course associated with any major change in farming practice, landowners and farmers require substantial support if they are to make this transition successfully. Therefore, revenue streams from both agricultural commodities and non-commodity goods, amenities and services are likely to be generally necessary to support transition. For example, some form of payment for environmental services or other similar revenue appears crucial to incentivize production of perennial biomass feedstocks for bioenergy in the Upper Midwest; analysis suggests that current prices offered for biomass per se are too low to attract landowners (John and Watson, 2007).

Third, new MFA enterprises must overcome inertia – both social and biophysical – that tends to reinforce established production and management systems. Factors militating against MFA include the opposition of dominant political actors, and long lag times before perennial-based MFA begins to produce commodities and ecological services. The virtuous circle model provides a device that can organize and direct strategic collective action to address such barriers (e.g. 'broken' links in the feedback loops of the virtuous circle). The sheer complexity of these situations makes tools for critical and systemic thinking vitally important to ACM groups attempting to establish MFA. Indeed, the virtuous circle model – or any systemic modelling framework – is principally useful for its heuristic value. For example, efforts to improve agricultural effects on water quality in the Chesapeake Bay have had limited success and the strategic basis of such efforts is certainly in need of critical and systemic rethinking (Diaz and Rosenberg, 2008).

Finally, the model underscores the need for development of new 'management regimes' for MFA that will support an ongoing process of intercoordinated 'knowledge innovation' across social, technical, market and policy sectors (Holtz et al., 2008). For example, such innovation is urgently needed to organize systems that pay farmers for production of environmental services. Such management regimes will rarely emerge spontaneously; rather, new levels of multistakeholder cooperation, communication, learning, deliberation and negotiation of conflicting interests are needed, requiring new coordinating mechanisms and agents (Pahl-Wostl, 2007). At present, we find very little appreciation of the need for new management regimes, coordinating mechanisms and agents among relevant stakeholder groups and agencies.

Despite these barriers, a considerable number of cases demonstrate the operation of virtuous circles (Selman and Knight, 2006; Steyaert and Jiggins, 2007). Most are in the socioeconomic, cultural and biophysical context of Western Europe, but in North America, state and local policy innovations in the northeastern region that preserve traditional agricultural landscapes in areas of rapid land-use change, and regional food system development around Toronto (Batie, 2003; Friedmann, 2007). Current examples are relatively limited in scope and scale, but suggest considerable potential for scaling up these strategies. Of course, since the virtuous circle process is driven by positive feedback, it is subject to various forms of self-limitation (e.g. depletion of suitable land area, saturation of demand for certain goods or services).

Subsystem level: collaborative social learning for multifunctional agriculture

MFA requires new forms of knowledge production. Typical agricultural research and extension systems emphasize technology transfer to increase commodity production. These systems are poorly designed for generating site-specific, agroecological knowledge for multifunctionality, or helping diverse stakeholders reconcile multiple goals for managing working landscapes. Sustainable enterprise development requires a more integral agricultural resource management strategy to yield multiple benefit streams. This demands balancing and synthesizing multiple socioeconomic goals – held by diverse individuals and institutions – within the biophysical constraints of specific agroecosystems. Typical agricultural research and extension institutions are poorly designed for negotiated and multidisciplinary activities to achieve multiple social goals (Warner, 2008). Different types of biological and practical knowledge must be coordinated and integrated to generate a multifunctional benefit stream. Advancing multifunctionality depends upon social learning, which we define as participatory research by diverse stakeholders to manage specific agroecosystems.

Agroecological partnerships have emerged around the USA to facilitate social learning about multifunctionality; partnerships have focused on horticultural crops, small grains and integrated pasture dairy farming (Warner, 2007a). New knowledge emerges from partnerships by co-ordination of social learning among agricultural producers, scientists in a range of disciplines, professional consultants, public agency officials and other parties possessing relevant knowledge (Warner, 2008). Partnerships incubate innovative practices and integrated approaches for diversifying commodity production and enhancing ecosystem services such as biodiversity conservation, watershed protection, amenity values or carbon storage (Steyaert and Jiggins, 2007).

Recent land-use conflicts in California provide a compelling example of the power of partnerships to increase multifunctionality. In these situations, partnerships evolved to cope with environmental regulatory pressure in farming. These pressures arose as high-value wine-grape production and high-value residential development have come into intimate contact in California, and heated public debate has erupted about pesticide and water use (Warner, 2007b). In response, growers, scientists and public officials have created new social networks to produce knowledge needed to reduce pesticide and water use by increasing the multifunctionality of wine-grape production systems. These developments have significantly reduced land-use conflict in a number of cases.

Agroecological partnerships also address knowledge gaps about multifunctional systems that can hinder enterprise development. For example, divergent views among stakeholders recently blocked a proposed programme to publicly subsidize perennial biofuel production systems in Minnesota. Conservation interests preferred species-rich perennial biomass production systems, while farmers and biomass processors preferred less-diverse systems. A well-designed perennial biofuel production system likely could meet key concerns of both parties. Such a system will require a careful and locally specific approach to design and management. Our experience suggests that a close coupling of research activity to multistakeholder processes of learning and deliberation can help resolve such stakeholder disputes. Partnerships can achieve this coupling, but only if certain key resources are available, so that individuals and institutions can facilitate shared research agendas, access resources and extend partnership networks along a 'vertical' dimension to influence research institutions and relevant public policy.

Supersystem level: re-visioning the social metabolism of American agriculture

Agriculture is the fundamental metabolic relationship between nature and society (FitzSimmons and Goodman, 1998). Early critics of agricultural industrialization addressed its environmental and human health consequences and laid the foundation for subsequent political and ethical critiques (Nestle, 2002). The popularity of recent bestsellers, such as *The Omnivore's Dilemma*, indicates broad public discontent with our agrifood system (Pollan, 2007).

Despite these stirrings, a thorough re-visioning of agriculture has only just begun. Studies in the political economy of the agrifood system indicate that transnational economic interests have essentially defined the popular understanding of the US food and agriculture system, and thus closed off consideration of alternative approaches to its configuration (Bonanno et al., 1994). To open up fresh ways of conceptualizing the metabolic relationship between agriculture and society, it will be necessary to re-design relationships within the agrifood system, and between this system and broader society. More specifically, increasing the multifunctionality of US agriculture depends crucially on marshalling public opinion and creating a much broader base of socioeconomic support for alternative rural development trajectories (Morgan et al., 2006; Warner, 2007b).

These developments depend in turn on broader and more intensive public engagement with the agrifood system. We contend that such engagement will require the emergence of a new 'imaginary', built on a premise of MFA and the interlinked processes of systemic change crucial to adoption of MFA. An imaginary is a multidimensional world view, a shared societal narrative incorporating cultural, social, economic and political dimensions. Imaginaries are powerful forces in public life, shaping public opinion and policy.

Our contemporary agrifood system discloses our collective imaginary about the environment, human health, social equity and the relationship between rural and urban life. Our society imagines these to be essentially discrete and at best loosely aggregated, if not completely separated (FitzSimmons and Goodman, 1998). The exclusive focus on commodity production has furthered the alienation of industrial agriculture from broader society. A new imaginary that highlights

MFA is needed because many key stakeholders do not yet recognize the potential of such an agriculture to address their concerns. For example, there are many concerns about animal agriculture for meat production in the USA, and there is some awareness that new grazing systems can address many of these concerns. However, from interviews with stakeholders in southeast Minnesota, we believe that relatively few advocates for change in animal agriculture and in associated agrifood systems are aware of the full range of beneficial environmental effects of well-managed grazing systems, and hence do not appreciate the potential size and strength of a coalition of support for grazing (Jordan et al. unpublished; Boody et al., 2005). Without significantly revision of an agricultural imaginary, MFA will continued to face scepticism and institutional inertia at the level of public judgment and policy formation.

In our theory of change, 'communities of ethical concern' play a crucial role in fostering a new imaginary, shared between broader society, rural America, and agricultural interests, and in driving change in public policy that shapes the agri-food system. They do this by educating various publics about the negative conse-quences of industrial agriculture, and by encouraging public support for MFA with sustainable rural enterprise development and policies that reward MFA. Communities of ethical concern have greatly increased in number and activity during the past decade, manifested by myriad citizen efforts to develop agrifood systems that balance economic development, social equity and environmental protection (Allen et al., 2003). Many projects work to promote a more 'civic agriculture', that is an agrifood system that is integrated into the social and eco-nomic development of a local community, and fosters participation in civil soci-ety (Lyson, 2004). These alternative, citizen imaginaries reveal a more integrated and ethical vision for America's agrifood system. Through changes in institu-tional and organizational behaviour and formation of cross-sector coalitions, significant political power can be mustered in support of policy changes neces-sary to support multifunctional agriculture and the operation of virtuous circles (Steyaert and Jiggins, 2007).

Applying the theory of change: the Koda Energy fuelshed project

A pilot project on sustainable bioenergy production is unfolding in the lower Minnesota River valley at the edge of the Minneapolis–St. Paul Metro region. Koda Energy LLC has invested $60 million to create a biomass-based co-generation system. The aim is to source 30% of fuel for the plant from local perennial energy crops, creating demand for ca. 50 000 tons per year of bio-mass, drawing from a 20-mile radius 'fuelshed'. If this biomass can be harvested from 10 000 acres of environmentally sensitive sites, major regional benefits could be realized on a very cost-effective basis; these include improved water quality, conservation of wildlife and other biodiversity, reduced flooding risks and greatly enhanced recreational and amenity value. Various stakeholder organ-izations are deploying resources for the Koda project, which is at present guided by an informal steering group (including co-author Jordan).

This group aims to create a regional 'multifunctional bioenergy fuelshed', a new sustainable land architecture for the joint production of regional economic and environmental benefits and renewable energy. As noted above, sustainable land architecture is the design and realization of complex land-use/ land-cover patterns that meet multiple human needs from multiple life-support systems while sustaining these systems for the long run (Turner et al., 2007). Through fuelshed development, the group hopes to help manage land-use change in this periurban region, in which many forces are exerting strong pressures on land use.

Land-use change is an issue of global importance, affecting food production, water resources, air quality, climate, biodiversity and infectious disease; better management of land-use change is certainly crucial to the sustainability of human societies (Foley et al., 2005; Turner et al., 2007). In particularly, better management of cumulative and multiscalar impacts of local land-use decisions on ecological service production is urgently needed in almost all arenas. Consider the case of water: intensification of land use during urban or agricultural development has many cumulative effects on water resources, exacerbating peak hydrological flows and soil erosion rates, and altering the quantity and quality of material and energy flows within surface and groundwater hydrologic regimes. The downstream consequences of these effects have impacts across many scales. These water-related cumulative effects of land-use change are expected to be particularly significant to societal well-being as humanity enters an era of water scarcity driven by population growth and climate change, and faces other novel problems such as flooding and erosion hazards resulting from changed precipitation patterns.

In response to such considerations, the project steering group is working to create and realize new MFA land architectures, applicable to the Fuelshed region and elsewhere, that can increase and sustain the production of multiple ecological services that are crucial to sustainable food, water and energy systems. The group is composed of students and research and extension faculty at the University of Minnesota–Twin Cities, staff in several state agencies, local government authorities and non-governmental organizations (NGOs). Guided by the theory of change articulated above, the steering group is working on a number of fronts. These efforts are described below, along with an assessment of progress in each area. Together, these analyses provide a detailed case study of an application and evaluation of the theory of change for MFA articulated above.

Enterprise development

At present, the Koda steering group is building foundations for a new enterprise–development framework. The first order of business is to create capacity for ACM. To do so, the group is synthesizing the following elements into a framework to support ACM, and working in close engagement with a multistakeholder community in order to apply, evaluate and refine this framework:

1   Methods for integration and communication of natural and social scientific knowledge that address the multiple functions of the landscape and assess the

wide-ranging and cumulative impacts of discrete decisions across spatial and temporal scales (Innes, 1998). These tools are needed to enhance learning among stakeholders in ACM. Relevant methods requiring integration include visualization and scenario analyses that facilitate realistic depictions of current and future conditions to effectively communicate the varied effects of decision alternative (Al-Kodmany, 1999; Hamilton et al., 2001; Sheppard, 2005). Especially important are tools for identifying and understanding social and biophysical stabilizing factors that limit or heighten the spatial and temporal influence of ACM (Pahl-Wostl and Hare, 2004; Holtz et al., 2008).

2   Processes for collaborative learning, decision making and collective action that bring natural and social science into effective interplay with other knowledge systems held by stakeholders. Such processes enable participants to jointly search for information, educate each other and engage in dialogue, and can produce better environmental outcomes and facilitate consensus (Friedmann, 1987; Innes, 1992; Bentrup, 2001; Margerum, 2002; Mandarano, 2008). Especially, we are testing the ability of land-use design processes as vehicles for learning, decision-making and action processes (Nassauer and Opdam, 2008).

To apply this framework, the steering group is now focusing on organizing work, aiming to identify and build interest in the project among multiple stakeholders, and to enrol stakeholders in fuelshed development ('interessment' and 'enrolment') (Steyaert and Jiggins, 2007). This is a slow and labour-intensive effort. It builds on extensive stakeholder interviews conducted in late 2007 (Evans et al. unpublished). These revealed broad and enthusiastic support for development of a multifunctional fuelshed, on the one hand, while also suggesting a need for development of both 'bonding' and 'bridging' social capital for fuelshed development. In particular, interviews revealed that stakeholders within broad categories were not collaborating or communicating with others within the same sector, nor across stakeholder sectors. Therefore, an important part of the steering group workplan is focused effort to create bonding ties between stakeholders within sectors. As well, efforts are needed to create bridging ties across stakeholder sectors.

To create bonding ties, the group recognizes three stakeholder sectors, and is seeking support to organize separate meetings for each of three groups: farmers and landowners, environment and policy organizations, and Shakopee community members. These categories are to act as guidelines for initiating communication about important issues. These meetings will improve communication and move toward building consensus between groups with similar interests and values on issues related to the Koda Project. The goal is to support participants in thinking critically and systemically about how the Koda project affects key interests and concerns of each sector, and to identify interest in further learning and collaboration within and across sectors. After within-sector meetings have occurred, we will organize and facilitate several cross-sectors meetings, which will be linked to a range of experiential learning opportunities (tours of the Koda Energy facility, tours of demonstration sites and other significant areas within the Fuelshed).

After these initial organizing efforts are complete, the steering group will continue to provide facilitation and support for a multistakeholder ACM network. Using an action research approach, we will draw on extensive experience with methods for systems learning (e.g. use of a range of tools that provide insight into social and biophysical systems that affect the Fuelshed and enterprise development). These include highly generalized discussions of 'vision' for landscape development, development of multiple scenarios for landscape development, participatory development of a range of social and biophysical component models that vary widely in scale, construction and purpose. In the application of these tools to support collaborative learning about the Fuelshed, and in deliberation about possible trajectories of Fuelshed design and development, it is crucial that participants view the methods and processes employed as salient (relevant to their interests), legitimate (unbiased and democratic) and credible (supportive of well-founded deliberation and decision making). An action research approach is essential, because such experience with ACM demonstrates that a highly eclectic, multimethod approach to both methods for systemic learning and for stakeholder deliberation and decision making are crucial (Steyaert and Jiggins, 2007; Voinov and Gaddis, 2008). Stakeholder observations and perceptions will be essential to evaluation of the framework for ACM that the steering group will create.

A second important dimension of enterprise development is providing opportunities for additional income streams to producers of perennial biomass crops. There is a willing buyer for large tonnages of perennial biomass from the fuelshed region; therefore, present work focuses on creation of a pilot system of payment for environmental services (PES), which appears crucial to profitable biomass production. We are working to create a brokering entity that can aggregate multiple 'environmental commodities' and other goods and services, market these and pay landowners. This effort is leveraging University of Minnesota and NGO research on local and regional demand for environmental commodities and other goods and services, which has found that a number of local and regional stakeholders are interested in providing PES in the Fuelshed. State and Federal policy measures now offer subsidy payments to 'biomass production areas' that have an existing market for biomass and in which sufficient biomass production will occur to support these markets. We are working to prepare a competitive application for these subsidies. To support these enterprise development efforts, we have organized a multistakeholder confederation of state agencies, NGOs, local partners, local and regional government officials, and University of Minnesota research and extension personnel. This group supports the learning, innovation and collective action that are essential to the operation of the virtuous circle model of enterprise development.

Agroecological partnership

University of Minnesota researchers have received substantial external grant support for research on the agroecology of multifunctional fuelshed landscapes in the Koda region. With these grants researchers are collaborating with other actors to enable the co-creation of localized knowledge. Co-researchers include local soil and water conservation district personnel, and other stakeholders such as Minnesota Department of Natural Resources (MN DNR) and the Nature Conservancy. MN DNR has recently catalysed the formation of a partnership by recognizing the value of perennial biomass agriculture to its conservation goals. The MN DNR is providing various financial incentives and helping to convene co-researchers around a common interest in the multifunctionality of the conservation area, and is an active co-researcher.

Key knowledge gaps that will be addressed by agroecological partnership include the following. First, spatial analyses are needed via geographic information system (GIS) to identify environmentally sensitive areas that are well suited to production of perennial biomass crops, and to identify opportunities to place biomass crops to create upland and low-land networks that can increase the effectiveness and cost-effectiveness of soil, water and biodiversity conservation in the Fuelshed. A range of agronomy research is needed on issues of establishment, harvest and management of perennial biomass crops. Wildlife management research is needed to increase wildlife/ biodiversity benefits from perennial biomass crops that are necessarily subject to harvesting and other forms of disturbance. Research is need to help guide placement of a range of biomass crops so as to support an economically rational flow of biomass to Koda Energy and other consumers of biomass from the Fuelshed. To do so, biomass crops with a range of harvest times will be needed to reduce the need for harvesting equipment and storage facilities. University of Minnesota research is now under way on biomass production, utilization, wildlife/ biodiversity conservation, green payments and hydrology/ water quality. Coordination with these projects so that their work supports – and is supported by – the work of the Koda agroecological partnership is both a major opportunity and a substantial challenge in organization and communication.

Re-shaping public opinion and policy

We are working to build support for policy changes that will support a multifunctional fuelshed, at local to regional levels. Local efforts concentrate on dialogue with local land-use and development planning agencies; state-level efforts emphasize environmental, agricultural and economic stakeholder groups, aiming to form a coalition that can win new state policy support on a number of levels. Finally, we are using strategic communications methods to reframe public awareness and understanding of bioenergy among key influencers of public opinion and policy.

Conclusions

Dramatic pressures are driving transformative changes in American agriculture. The socioecological environment is rapidly evolving, and thus challenging conventional farming practices and related systems of agriculture. Scientific, economic and social analysis indicates that a more multifunctional agriculture can reconcile the concerns and interests of environmental, economic and agricultural sectors of society. Realizing the potential of MFA poses multiple challenges, as it will depend upon co-ordinated scientific and social action, integrated across a multiple scales, sectors and systems, as we have outlined in the Koda Fuelshed case study. The theory of change outlined above offers an ecologically informed heuristic for negotiating these challenges. In our view, such heuristics are crucial to cross-sector initiatives for sustainable development such as the Koda Fuelshed project, which are now developing and refining strategies for increasing multifunctionality in agriculture and in land use generally.

References

Al-Kodmany, K. (1999). Using visualization techniques for enhancing public participation in planning and design: Process, implementation, and evaluation. *Landscape and Urban Planning*, **45**, 37–45.

Allen, P., FitzSimmons, M., Goodman, M., et al. (2003). Shifting plates in the agrifood landscape: The techtonics of alternative agrifood initiatives in California. *Journal of Rural Studies*, **19**, 61–75.

Argent, N., Smailes, P. and Griffin, T. (2007). The amenity complex: Towards a framework for analyzing and predicting the emergence of a multifunctional countryside in Australia. *Geographical Research*, **45**, 217–232.

Armitage, D., Berkes, F. and Doubleday, N. (eds) (2007). *Adaptive Co-management: Learning, Collaboration and Multi-Level Governance*. UBC Press, Vancouver.

Batie, S.S. (2003). The multifunctional attributes of Northeastern agriculture: a research agenda. *Agricultural and Resource Economics Review*, **32**, 1–8.

Bentrup, G. (2001). Evaluation of a collaborative model: a case study analysis of watershed planning in the Intermountain West. *Environmental Management*, **27**, 739–748.

Bonanno, A., Busch, L., Friedland, W.H., et al. (eds) (1994). *From Columbus to ConAgra: The Globalization of Agriculture and Food*. University Press of Kansas, Lawrence, KS.

Boody, G., Vondracek, B., Andow, D.A., et al. (2005). Multifunctional agriculture in the United States. *Bioscience*, **55**, 27–38.

Brown, R., Rosenberg, N., Hays, C., et al. (2000). Potential production and environmental effects of switchgrass and traditional crops under current and greenhouse-altered climate in the central United States: a simulation study. *Agriculture, Ecosystems and Environment*, **78**, 31.

Diaz, R. and Rosenberg, R. (2008). Spreading dead zones and consequences for marine ecosystems. *Science*, **321**, 926–929.

Donner, S. and Kucharik, C.J. (2008). Corn-based ethanol production compromises goal of reducing nitrogen export by the Mississippi River. *Proceedings of the National Academy of Sciences USA*, **105**, 4513–4518.

Eaglesham, A. (2006). *Proceedings of the Third Annual World Congress on Industrial Biotechnology and Bioprocessing*, Toronto, Canada 11–14 July 2006. Available at: http://nabc.cals.cornell.edu/pubs/WCIBB2006_proc.pdf (accessed August 2012).

FitzSimmons, M. and Goodman, D. (1998). Incorporating nature: Environmental narratives and the reproduction of food. In: *Remaking Reality: Nature at the Millennium* (eds B. Braun and N. Castree), pp. 194–220. Routledge Press, New York.

Flora, C.B. (ed.) (2001). Shifting agroecosystems and communities. In: *Interactions Between Agroecosystems and Rural Communities*, pp. 5–14. CRC Press, New York.

Foley, J., DeFries, R., Asner, G., et al. (2005). Global consequences of land use. *Science*, **309**, 570–574.

Friedmann, H. (2007). Scaling up: bringing public institutions and food service corporations into the project for a local, sustainable food system in Ontario. *Agriculture and Human Values*, **24**, 389–398.

Friedmann, J. (1987). *Planning in the Public Domain: From Knowledge to Action*. Princeton University Press, Princeton.

Gantzer, S.H., Anderson, A.L., Thompson, J.R., et al. (1990). Research reports: estimating soil erosion after 100 years of cropping on Sanborn Field. *Journal of Soil, Water and Conservation*, **45**, 641–644.

Goolsby, D.A., Battaglin, W.A., Lawrence, G.B., et al. (1999). *Flux and Sources of Nutrients in the Mississippi-Atchafalaya River Basin: Topic 3 Report for the Integrated Assessment of Hypoxia in the Gulf of Mexico*. NOAA Coastal Ocean Program, Silver Spring, MD.

Gunderson, L.H. and Holling, C.S. (2002). *Panarchy: Understanding Transformations in Human and Natural Systems*. Island Press, Washington, DC.

Hamilton, A., Trodd, N., Zhang, X., et al. (2001). Learning through visual systems to enhance the urban planning process. *Environment and Planning B*, **28**, 833–845.

Hanson, J.D., Liebig, M.A., Merrill, S.D., et al. (2007). Dynamic cropping systems: increasing adaptability amid an uncertain future. *Agronomy Journal*, **99**, 939–943.

Hey, D.L., Urban, S. and Kostel, J.A. (2005). Nutrient farming: the business of environmental management. *Ecological Engineering*, **24**, 279–287.

Holtz, G., Brugnach, M. and Pahl-Wostl, C. (2008). Specifying "regime": a framework for defining and describing regimes in transition research. *Technological Forecasting and Social Change*, **75**, 623–643.

Innes, J. (1992). Group processes and the social construction of growth management: Florida, Vermont, and New Jersey. *Journal of the American Planning Association*, **58**, 440–453.

Innes, J. (1998). Information in collaborative planning. *Journal of the American Planning Association*, **64**, 52–63.

John, S. and Watson, A. (2007). *Establishing a Grass Energy Crop Market in the Decatur Area*. Agricultural Watershed Institute, Decatur. Available at: http://www.agwatershed. org (accessed January 2009).

Jordan, N., Boody, G., Broussard, W., et al. (2007). Sustainable development of the agricultural bio-economy. *Science*, **316**, 1570–1571.

Jorgensen, U., Dalgaard, T. and Kristensen, E.S. (2005). Mass energy in organic farming: the potential role of short rotation coppice. *Biomass and Bioenergy*, **28**, 237–248.

Kleijn, D., Baquero, R.A., Clough, Y., et al. (2006). Mixed biodiversity benefits of agri-environment schemes in five European countries. *Ecology Letters*, **9**, 243–254.

Lyson, T.A. (2004). *Civic Agriculture: Reconnecting Farm, Food, and Community*. University Press of New England, Boston.

Magner, J.A., Payne, G.A. and Steffen, L.J. (2004). Drainage effects on stream nitrate-N and hydrology in south-central Minnesota (USA). *Environmental Monitoring and Assessment*, **91**, 183–198.

Mandarano, L. (2008). Evaluating collaborative environmental planning outputs and outcomes: restoring and protecting habitat in New York–New Jersey Harbor Estuary Program. *Journal of Planning Education and Research*, **27**, 456–468.

Margerum, R. (2002). Evaluating collaborative planning: implications from an empirical analysis of growth management. *Journal of the American Planning Association*, **68**, 179–193.

Meyer, B.C., Phillips, A. and Annett, S. (2008). Optimizing rural land health: from landscape policy to community land use decision-making. *Landscape Research*, **33**, 181–196.

Morgan, K., Marsden, T. and Murdoch J. (2006). *Worlds of Food: Power, Place and Provenance in the Food Chain*. Oxford University Press, London.

Nassauer, J.I. and Opdam, P. (2008). Design in science: extending the landscape ecology paradigm. *Landscape Ecology*, **23**, 633–644.

Nestle, M. (2002). *Food Politics: How the Food Industry Influences Nutrition and Health*. University of California Press, Berkeley, CA.

Pahl-Wostl, C. (2007). The implications of complexity for integrated resources management. *Environmental Modeling and Software*, **22**, 561–569.

Pahl-Wostl, C. and Hare, M. (2004). Process of social learning in integrated resource management. *Journal of Community and Applied Social Psychology*, **14**, 193–206.

Plummer, R. and Armitage, D. (2007). A resilience-based framework for evaluating adaptive co-management: linking ecology, economics and society in a complex world. *Ecological Economics*, **61**, 62–74.

Pollan, M. (2007). *The Omnivore's Dilemma: A Natural History of Four Meals*. Penguin, New York.

Robertson, G.P., Paul, E.A. and Harwood, R.R. (2000). Greenhouse gases in intensive agriculture: contributions of individual gases to the radiative forcing of the atmosphere. *Science*, **289**, 1922–1925.

Selman, P. and Knight, M. (2006). On the nature of virtuous change in cultural landscapes: exploring sustainability through qualitative models. *Landscape Research*, **31**, 295–307.

Sheppard, S. (2005). Landscape visualization and climate change: the potential for influencing perceptions and behavior. *Environmental Science and Policy*, **8**, 637–654.

Steyaert, P. and Jiggins, J. (2007). Governance of complex environmental situations through social learning: a synthesis of SLIM's lessons for research, policy and practice. *Environmental Science and Policy*, **10**, 575–586.

Tilman, D., Hill, J. and Lehman, C. (2006). Carbon-negative biofuels from low-input high-diversity grassland biomass. *Science*, **314**, 1598–1600.

Turner, L., Lambin, F. and Reenberg, A. (2007). The emergence of land change science for global environmental change and sustainability. *Proceedings of the National Academy of Sciences USA*, **104**, 20666–20671.

Voinov, A. and Gaddis, E. (2008). Lessons for successful participatory watershed modeling: a perspective from modeling practitioners. *Ecological Modelling*, **216**, 197–207.

Warner, K.D. (2008). Agroecology as participatory science: emerging alternatives to technology transfer extension practice. *Science, Technology and Human Values*, **33**, 754–777.

Warner, K.K. (2007a). *Agroecology in Action: Extending Alternative Agriculture Through Social Networks*. MIT Press, Cambridge.

Warner, K.K. (2007b). The quality of sustainability: agroecological partnerships and the geographic branding of California winegrapes. *Journal of Rural Studies*, **23**, 142–155.

Zhang, W., Ricketts, T.H., Kremen, C., et al. (2007). Ecosystem services and dis-services to agriculture. *Ecological Economics*, **64**, 253–260.

10

Supply Chain Management and the Delivery of Ecosystems Services in Manufacturing

Mary Haropoulou,[1] Clive Smallman[2] and Jack Radford[3]

[1] Lincoln University, Canterbury, New Zealand and the University of Western Sydney, Australia
[2] University of Western Sydney, Australia
[3] Lincoln University, Canterbury, New Zealand

Abstract

Supply chain management is fundamental to conventional models of economic development, but is conventionally associated with the optimized movement of raw materials and finished products across a network of interconnected and interacting firms. The ecological impact of supply chains is acknowledged in the development of the knowledge around so-called sustainable supply chains. However, the degree of integration of the concept of ecosystems services and wider ecological economic models in supply chain management theory is apparently limited.

We first review models of conventional and ecological economic systems and of varying conceptualizations of supply chain management. From this base we offer a fresh synthesis of sustainable supply chain management and the concept of an ecological economic system. We then evaluate the validity of this synthesis in describing and then discussing a qualitative case study of woollen carpet yarn manufacture, grounded in a life-cycle assessment of the production process and a 9-month field study of decision making in product creation, with a special interest in sustainable outcomes. We find evidence to suggest that our conceptualization may offer a route to improving life cycle assessment as a means of analysing the sustainability of a supply chain.

Ecosystem Services in Agricultural and Urban Landscapes, First Edition. Edited by Steve Wratten, Harpinder Sandhu, Ross Cullen and Robert Costanza.
© 2013 Commonwealth Scientific and Industrial Research Organisation.
Published 2013 by John Wiley & Sons, Ltd.

Towards the sustainable economic production of goods and services?

Conventional supply chain management (SCM) may be defined as the management of a network of interconnected firms interacting to provide products and services required by clients or customers. SCM therefore is fundamental to conventional models of economic development. Conventional supply chain managers seek optimal supply chain performance as a means of heightening economic performance and growth through increased efficiency. Typically, the pursuit of 'efficiency' rarely accounts for the 'natural environment', although there are some positive signs that business is beginning to take note of the importance of issues such as this (Staib, 2009). However, the term 'ecosystem services' is not one that we believe conventionally trained supply chain managers will be familiar with. This assertion is based on our reading of the literature on organization and management theory and our collective experiences in manufacturing (particularly of wool yarn) and service industries.

That stated, there has been increased awareness of the importance of 'greening' the supply chain. This is attributable not just to the ongoing pressure from governments, regulatory bodies and communities for companies to minimize their environmental impact (Zhu et al., 2005; Vachon and Klassen, 2006). It is also attributable to issues such as global warming, reductions in air quality, the pollution of waterways and the loss of biodiversity, which are arguably attributable, to varying extents, to the coordinated activities of firms in a supply chain. However, even here, environmental degradation is viewed as an economic inefficiency, with the focus still firmly on conventional economic growth (van Hoek, 1999).

The question we address in this chapter is: how can we move organizations to recognize and account for the provision of ecosystems services in the production of goods and services?

We provide some answers on the basis of a review of theories of ecological economics and supply chain management, which we synthesize as 'sustainable supply chain management'. We then evaluate this synthesis in examining a case study of woollen carpet yarn manufacture.

Ecological economics and supply chain management: a review and synthesis

Conventional economic and ecologically economic production

Measured by gross national product (GNP), conventional economic production combines manufactured capital (e.g. infrastructure, plant), human capital (labour) and natural capital (land) to develop goods and services (Fig. 10.1). Governed by economic regulation, the consumption of goods and services produces individual utility. The growth of GNP requires investment in building or maintaining manufactured capital, learning or research to improve labour productivity or the improvement of land. Each primary factor can, within limits, be used to substitute for each other. Hence, damage to 'land' can be compensated for by increased use of other inputs. Property rights are simplistic (Costanza, 2000).

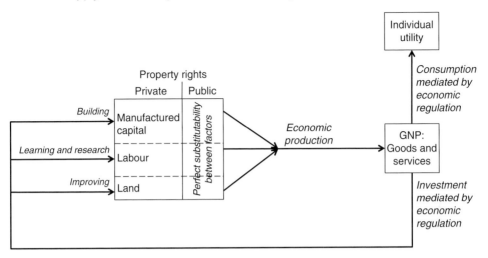

Fig. 10.1   Conventional economic system (from Costanza, R. (2000) Social goals and the valuation of ecosystem services. *Ecosystems*, **3**, 4–10).

Reflecting real-world complexity in economic production remains the foundation of ecological economic systems (see Fig. 10.2). It remains focused upon the production of goods and services, and in aggregate GNP, but inputs, outputs and interactions of the models are more complex. Rather than 'simple' land the inputs sourced from the natural world (e.g. the products of ecosystems, such as minerals) are defined as natural capital. These are combined with human capital (individual or pooled knowledge, experience, skills and physical effort), social capital (e.g. interpersonal connections, institutional arrangements, rules and norms governing interactions) and manufactured capital in economic production. Natural capital 'hosts' the ecological services and amenities (ecosystems services) that impact community and well-being, in the pursuit of which the act of living combines each of the various forms of capital. Well-being positively affects human capital. The ownership of the various forms of capital is complex and they are not easily substituted (Costanza, 2000).

Economic production also produces well-being and waste, the former through the provision of livelihoods and meaningful work. So too will the regulated consumption of goods and services produce well-being, although it may also produce waste. Waste itself will emit heat into the ecosystem and may negatively impact ecosystems services, and each form of capital. GNP may arise through restoring and conserving natural capital, in developing human capital, in governing civil society in pursuit of social capital, and in building or maintaining manufactured capital (Costanza, 2000).

What it is important to recognize is that both models rely heavily on the 'economic production system'. This production system is a chain of subsystems that extends from the extraction of raw materials through production processing, marketing and sales, and eventually to disposal or recycling (Caldwell and Smallman, 1996). In conventional economics this chain is concerned with 'adding' economic value as raw materials are transformed to finished goods. Hence, in its original formulation, the value chain's focus is on the efficient

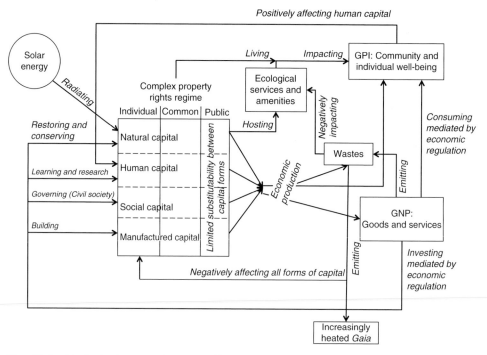

Fig. 10.2   Ecological economic system (from Costanza, R. (2000) Social goals and the valuation of ecosystem services. *Ecosystems*, **3**, 4–10).

supply of raw materials to efficiently manufacture goods or to efficiently produce services (Porter, 1985) through optimizing an organization's primary activities (e.g. inbound logistics, operations, outbound logistics, marketing and sales, service). Under ecological economics, the focus of the chain is on the sustainable production or delivery of goods or services, which requires not just optimization of primary activities, but also considered approaches to secondary functions (e.g. procurement, information and production technology services, human resource management, firm infrastructure services, corporate governance and executive management). Hence, enabling a truly ecologically economic approach to production turns on the transformation of the structure and nature of this system.

Conventional SCM: economic efficiency through distribution network configuration and strategy

The modern supply chain originated in the development of the canal systems and railways of the industrial revolution in Europe, in the opening up of the American west and in the reconstruction of Western Europe and Japan following the Second World War. SCM emerged as a business philosophy in the 1980s as organizations rode the first wave of flexible specialization. Conventional SCM involves the management of materials, information and financial flows within a network of suppliers, manufacturers, distributors and end customers, integrating the dependent activities, actors and resources between the different levels of the points of origin and consumption in channels (Svensson, 2007). The coordination and integration of these flows within and across companies should enhance various companies' competitive advantage (Simchi-Levi et al., 2004).

Green SCM: the economic inefficiency of waste

The majority of firms who claim to have green supply chains, make their claim on the basis of production efficiencies that reduce waste at key points in production (Bowen et al., 2001), usually driven by regulatory requirements. It is arguable that 'greening' the supply chain is a much broader concept than just 'waste reduction', encompassing: the sustainable sourcing of raw materials (e.g. not depleting naturally occurring minerals); the use of power used in production from renewable sources (instead of fossil fuels); the assurance of efficient logistics (e.g. optimizing fuel usage), inter alia. However, the foundation of all of these best practices is the need to conserve resources (e.g. fossil fuels, natural forests, mineral resources) whilst not compromising the commons (water, air, land). Hence, acknowledging, as we do shortly, the breadth of the concept of green SCM (GSCM), we argue strongly that, at an operational level, conventional GSCM comes down to the limitation of waste.

The conventional GSCM argument runs that applying an environmental perspective in the daily operations of a firm makes good business sense as it helps improve their performance and hence their productivity. Greening the supply chain in this manner (Caldwell and Smallman, 1996) includes a wide variety of activities and an increased degree of co-operation between firms in order to minimize the logistical impact of the material and information gathering flows (Vachon and Klassen, 2006). However, green supply extends beyond the logistics area, and may impact the purchasing function where attention is on recycle, reuse and reduce (Min and Galle, 2001). Hence, green supply practices encompass activities both internal to an organization as well as external, whether related to preventing pollution before it is generated, recycling waste, extracting resources and raw materials, or capturing harmful pollutants followed by proper disposal (Vachon and Klassen, 2006).

Maintaining a green supply chain can be achieved through integration of environmental thinking into the whole of the supply chain, ranging from material sourcing and selection to product design and manufacturing, all the way to consumer delivery of the final product as well as managing the end of product life (Srivastava, 2007). Managing the end of life of a product extends the one-way supply chain to construct a semi-closed loop that includes product and packaging recycling, re-use and remanufacturing operations (Beamon, 1999). Handfield et al. (2005) state that environmental SCM involves introducing and integrating environmental issues and concerns into SCM processes by auditing and assessing suppliers on environmental performance metrics issues.

However credible this approach is, it still revolves around the conventional economic arguments that production inefficiencies are wasteful, and that conventional economic growth offers the best means of improving the natural environment. In this paradigm environmental waste is a barrier to the efficient generation of GNP, negatively impacts the natural environment and ultimately reduces individual utility or welfare. It retains an overly simplistic view of the interaction between humans, the economic production system and ecosystems. It reflects neither the role that ecosystems play in our lives nor their centrality to the economy. In other words, it creates problems for the natural environment, and fails to deal with social issues that offer an equally valid set of challenges for firms.

Sustainable SCM: connecting social, economic and ecological performance

In an ecological economic system we see a much more complex chain of production that acknowledges waste as important, but also recognizes the complexity of the supply chain through the representation of feedback throughout such a system. The key difference in this approach is that waste is not only about production inefficiency; it is about wide-scale impact on ecology and all forms of capital that play a role in human and other life. Especially important is the concept of ecological services or amenities upon which life relies. What this more complex model of economy does is to reconnect the supply chain with the ecosystem in which it operates. Hence, the run-down of finite raw materials stocks, the escalating deterioration of climate, overflowing waste sites, increasing pollution (Srivastava, 2007) and impending energy shortages (Hartmann, 2004 cited in Beamon, 2008) are each fully linked to SCM. Moreover, the importance of human well-being is brought to the fore.

The ecological economic model draws together economic (e.g. manufactured capital, GNP, goods and services), ecological (e.g. natural capital, wastes, ecological services) and social (e.g. human capital, social capital, well-being) performance (Costanza, 2000; Carter and Rogers, 2008). As such it implicitly reflects Triple Bottom Line Theory (Seuring and Müller, 2008) through a definition of sustainable SCM (SSCM) that draws upon a thorough analysis of the sustainability and SCM literatures:

> the strategic, transparent integration and achievement of an organization's social, ecological, and economic goals in the systematic coordination of key inter-organizational business processes for improving the long-term economic performance of the individual company and its supply chains.
>
> Carter and Rogers (2008, p. 368)

Whilst the green supply chain lies at the intersection of economic performance and ecological performance, it is largely concerned with improved risk management, contingency planning, supply reliability and logistics. Genuine sustainability lies at the union of economic, ecological and social performance, adding improved transparency, and stakeholder engagement and supplier relationships to the mix.

So, through creating sustainable processes firms may still generate commercial benefits without impacting heavily on the environment (Srivastava, 2007), and may achieve social goals too. Moreover, several studies advocate the positive correlation between sustainable practices and economic performance (Rao and Holt,

2005; Zhu et al., 2005; Vachon and Klassen, 2006; Seuring and Müller, 2008). From the selection of environmentally friendly raw materials through to manufacturing, packaging, storage and distribution sustainable practices can really enhance and impact positively on the environment (van Hoek, 1999). Improvements are not only limited to product-related changes or to manufacturers; supplier selection and management is also important and can extend changes upstream in the supply chain (Bowen et al., 2001). Over the last decade or so, SSCM has emerged as an important component of the strategies of a number of companies (Lee, 2008).

This evolving approach to sustainability in business is at least in part due to a heightened awareness among consumers and individuals of basic sustainability issues. Communities around the world are more than ever demanding cleaner air, water and soil (Beamon, 2008), in the context of the protection of human rights. Consumers are not necessarily in pursuit of a low-cost product, and base their purchase decisions partly upon the perceived sustainability attributes that a product or service poses (Beamon, 2008). In fact, according to Drumwright (1994), 80% of consumers agree to pay more for sustainable products. These consumer decisions exert pressure on companies to develop more proactive programmes and take initiatives to develop and implement sustainable strategies that, as well as preserving the environment and promoting well-being, will enhance their efficiency and effectiveness (Carter and Rogers, 2008).

Bringing SSCM to organizations requires a different approach across the value chain. Instead of a primary focus on logistics and technology, each primary activity and secondary function must take a role in the achievement of sustainability goals. Given the direction in which the world (natural, economic and social) is evolving, organizations may have little choice other than to adopt this holistic approach, overcoming the primacy of production or marketing in pursuit of margin. Instead, the requirement is to focus on the production of wealth, in its broadest sense of well-being. However, the model of SSCM presented by Carter and Rogers (2008) provides only a definition of the key concepts of SSCM; it does not locate these components in the production process, and fails to provide a system-wide view of sustainability in the supply chain and how this enables an ecological economic model of production.

Enabling ecological economics: SSCM

SSCM has a natural fit with the ecological economic model of production in that each of its elements influences or is influenced by inputs and outputs of the ecological economic production process (Fig. 10.3). This fit allows us to propose a synthesis of how SSCM facilitates the development of ecological economics.

Through economic production, facilitated by a supply chain, natural capital is transformed into goods and services (and waste) and enabling, community and individual well-being. Stocks of natural capital may increase if it is restored or conserved through private or state action. Natural capital hosts ecological services and amenities (ecosystems services), which impact community and individual well-being, and which in turn may be negatively impacted by waste. Waste may also lead to global warming and may negatively affect natural capital.

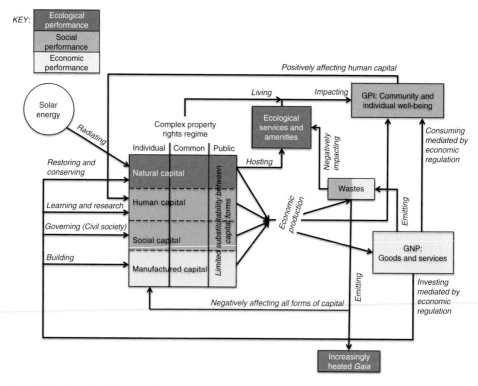

Fig. 10.3   Sustainable supply chain management and ecological economics (from Costanza, R. (2000) Social goals and the valuation of ecosystem services. *Ecosystems*, **3**, 4–10, with additional information from Carter and Rogers (2008)).

This is in essence one-half of the conventional view of sustainable supply chains or as we have portrayed them 'green' supply chains. Sourcing of raw materials, inbound and outbound logistics, and efficient and effective manufacture are each areas where environmental issues are commonly considered by organizations. However, we contend that too often they are considered purely within the boundaries of the firm. Rarely is the systemic view that we have synthesized considered – especially in terms of ecosystems services. This conventional view also tends to downplay the other vital elements of sustainable production.

Facilitated by SCM through economic production, human and social capital is transformed into goods and services, community and individual well-being and wastes. Positive returns to human capital are made through the positive affect of community and individual well-being, and through investment in learning and research that is mediated by economic regulation. Human capital may be negatively affected by waste (e.g. value added lost in production waste is not available for investment in human capital). Positive returns to social capital are made through investment in governing a 'civil society' that is mediated by economic regulation.

There is a common misconception of SCM purely as the management of transport and logistics. Another perspective dresses SCM purely as a branch of marketing. However, under the current proposition, we see SSCM as something deeper than these simple definitions. A genuinely sustainable supply chain is a reflection not just of its environmental 'credentials' as suggested in our first proposition, but also its human and social elements. SSCM requires highly skilled and committed individuals to interact in a highly professional manner to assure efficient and effective supply of raw materials, goods and services. For supply chains to work sustainably, there needs to be investment in individual knowledge and skills, but there also has to be investment in the social networks that are arguably the essence of supply chains.

SCM facilitates, through economic production, the transformation of manufactured capital into goods and services, community and individual well-being and wastes. Return on manufactured capital is made by investment in building more capital, mediated by economic regulation. Manufactured capital may be affected by waste (because of a misallocation of investment that might have been increased manufactured capital).

In one sense this is a reflection of where SCM started out from – the efficient use of manufacture to supply goods and services, yet it is also the other 'half' of what has become conventional or 'green' SCM.

In organizations, and particularly in commercial organizations, SSCM is a common thread that has the potential to tie ecological, social and economic performance elements together. Without this holistic approach to managing the economic production process, it will be difficult to fully operationalize the ecological economic model with its goal of sustainable human well-being. Moreover, ecological economics strengthens considerably the theoretical base for SSCM, enhancing its definition by demonstrating that the breadth of its influence across the processes that comprise an ecologically economic production system.

A case in point: 'what do we do with it now?'

We now turn to a short case study with which to evaluate our model of SSCM. The study grew out of conversations between all three present authors. Of these, Jack Radford is owner–director of a wool yarn manufacturing company, which we shall call WYM. Unusually, for a small business owner (in the experience of the other authors), Jack is interested in business sustainability (in all of its meanings – ecological, financial and social) and invited the other authors to study WYM's manufacturing process, with a view to addressing issues around its ecological footprint through developing a life-cycle assessment (LCA).[1,2]

---

[1] Mary Haropoulou was an 'embedded' participant observer at WYM from November 2009 to October 2010. Her early experiences of the firm were in the development of an LCA for WYM. Subsequently, she began observing the firm in general and attending board and management (formal and informal) meetings. Mary recorded meetings when allowed to and has transcribed much of the data collected thus. She also developed ethnographic field notes as she observed and later as she transcribed or listened to her digital recordings. Further data were collected in the form of official company documents (e.g. process manuals and annual reports). We analysed the data using a conventional qualitative data 'coding' approach (Miles and Huberman, 1994) in which we coded for the key terms used in our model of SSCM and ecological economics. We employed NVivo9 qualitative data analysis software in order to optimize coding. The reliability of our coding was checked through conventional check coding of samples of transcripts by colleagues at both Lincoln University and the University of Western Sydney.
[2] Whilst not completely germane to the focus of this chapter, based on our analysis of a large data set, one major issue for us lies in the manner in which product life cycles are assessed. LCA is mainly used for

Part way through what was a relatively simple LCA, Jack asked two questions: 'what are we going to do with it when you finish' and 'how can we learn from it and make the business better than it is?' When we began to probe the underlying logic of his questions, Jack laid out an issue he felt was common amongst industrialists. He observed that he and others believe that much of the analysis done in the name of sustainability – or to be more accurate corporate environmentalism –'stops' at the point when an LCA report is issued, and in essence what follows is 'greenwash',[3] implicitly supporting the dominance of conventional or, at best, green SCM. Jack viewed the LCA as the start point of a much grander project, one in which the assessment fed in to the development of a new sustainable strategy for the firm and ultimately the wool carpet industry. Hence, the chance to observe the operations of WYM seemed to us to offer an opportunity to explore just how realistic the adoption of our model of SSCM is.

WYM background

WYM commenced business in 1992. From the outset, it concentrated on producing high-quality innovative solutions for the apparel textile market using the 'felted yarn' technology. During 2000, there was a noticeable change from hand knitted apparel yarn to carpet and rug yarn, following a slump of interest in hand knitting. The company needed a new direction in order to remain in business. A long-time textile investor became WYM's angel investor and helped the company in times of need by investing in the business and directing energy towards the manufacture of specialized rug and carpet yarns, supplying mostly upmarket custom carpet and rug makers. WYM now occupies a position as a premium yarn supplier to major carpet manufacturers in New Zealand, the USA and Europe, who operate mainly in the corporate carpet market.

In 2006, due to their increase in production they moved premises to their present location (on New Zealand's South Island) with plenty of room to establish a facility capable of handling the large customer volumes required and the diverse range of products the company now produces for top-end carpet and rug makers globally. WYM claim to be a supplier of choice to 80% of their customers.

---

comparing the environmental impacts of products and not for evaluation (Curran, 2008). LCA aims at providing a comprehensive view of environmental impacts, but not all types of impacts are equally well covered in a typical LCA (e.g. impact assessments of land use, impacts on biodiversity and resource aspects such as freshwater resources (Finnveden et al., 2009)). Moreover, Carmody and Trusty (2005) argue that LCA does not easily handle such issues as uncertainty, risk related to toxic releases and site-specific resource extraction effects. Finnveden and Ekvall (1998) argue that an LCA does not produce the information that is envisaged by the ambitious LCA definition and seldom are the results of an LCA easy to reproduce. Jeswani et al. (2010) find that LCA has matured over several decades, becoming part of the broader field of sustainability assessment. To strengthen LCA as a tool and increase its usefulness in decision making around sustainability, they contend that the ISO LCA framework needs developing to integrate and connect with other concepts and methods.

[3] 'Greenwash' is defined as public relations releases or marketing efforts that are deceptively used to promote the perception that a company's policies or products have a lower ecological impact than is actually the case (Strasser, 2011).

One of the founders background in science and textiles gave the company a solid foundation for the hand-knitted yarn industry. His expertise from being a member of the New Zealand Wool Board, a miniature prototype of a 'felting machine' (that he many years ago brought back home in his suitcase from a visit to England's wool yarn heartland in West Yorkshire), coupled with his belief in felted yarn technology, are the foundations of the company's current strategy. Felted yarn differs from the spun wool that is traditionally used in carpets in that the wool fibres are deliberately shrunk and tied together to improving the strength and durability of the yarn. This results in a highly differentiated product that carries unique characteristics sought after by manufacturers of premium-quality carpets and rugs. However, the felting process that achieves this treatment uses heated water, detergents and dispersants, all of which pose challenges to ecological sustainability. The process as a whole is energy intensive.

The company is renowned internationally as a leader in the area of felted yarn technology with a focus on producing products that are varied, technically demanding, well styled and difficult for competitors to copy, with a very low or zero environmental impact. This positioning is reflected in all parts of the business from technically sophisticated enthusiastic staff, to market identification and distinctive branding. An international yarn innovation award, and a national supreme award for excellence and innovation recognized WYM's reputation as an innovator in technological and sustainable development in energy efficiency and renewable energy.

The economic production of wool yarn

At WYM the production system transforms raw wool into felted carpet and rug yarn for export (Fig. 10.4) through:

1 **Opening and blending**. Wool bales arrive at the factory, are opened and the wool fibres are broken up. This mechanical process employs the moisture content of the fibre, added water and a water-soluble oil to break up the wool clumps and makes the fibres more manageable for the next process.
2 **Carding**. The wool fibres are aligned together so that they are more or less parallel to each other to produce a wool sliver.
3 **Combing**. Only employed if Merino wool is being processed, this removes any vegetable matter and short fibres that remain after carding.
4 **Gilling** is a further step into making the wool sliver consistent and even, by further straightening the wool fibres.
5 **Roving** converts the wool sliver to a lightly spun yarn.
6 **Felting** is the heart of WYM's process and is a significant point of difference in the yarn that they produce. Hot water and detergents are used to rapidly shrink the yarn, producing its felted appearance.
7 **Drying** the yarn is crucial to achieve the moisture content required by customers.
8 In **winding** the felted yarn is wound on to cones ready for packaging. Alternatively, the factory may produce **hanks** (a specific length or weight of coiled yarn), which may be dyed (known as sample dying) or simply packaged.

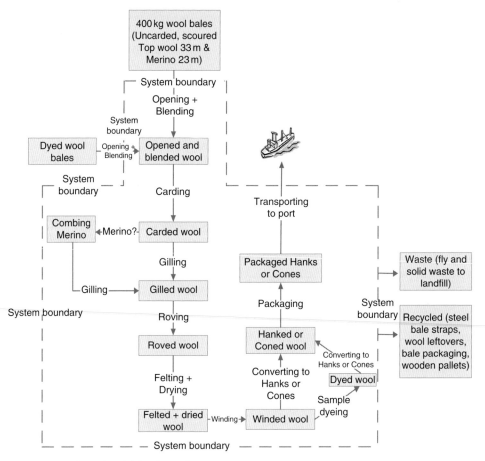

Fig. 10.4   Life cycle of felted wool yarn.

9   **Packaging** prepares the hanks or cones on pallets or in bales for shipping by container.
10   Pallets or bales are shipped by road container to the local port.

Goods

The focus of the company is on producing yarns that are varied, technically demanding, well styled and difficult for competitors to copy. WYM's yarns are manufactured to customer specifications and WYM can also manufacture similar products on demand and a small amount of non-felted yarns. WYM produces felted wool carpet yarn on cones or in hanks for major carpet manufacturers in Europe, the USA and New Zealand. A limited amount is dyed.

On average about 45 tonnes of high-quality felted yarn a month (75% of total revenue) exported to leading rug and carpet manufacturers in European markets (Netherlands, Germany), USA, Hong Kong and China. Australian and New Zealand manufacturers take up the remaining 25%.

Wastes

Wool waste is recycled through a recycling company that uses the waste as a raw material to manufacture environmentally sustainable products such as building insulation, horticulture goods, protective covers for horses and lambs, commercial furniture removal blankets and others. On average over the year October 2008–September 2009, in each month, WYM handed over 1288 kg of wool waste to the recycler. This could easily have been transported to landfill (at a cost) and so reduces the ecological impact of WYM's activities.

Water used to wash, clean and dye the wool is discharged as an effluent in the local council waste water stream. A limited amount of solid waste that ends up in landfill comprised of wool dust (fly) as well as other office or factory waste.

Ecological services and amenities

It is fair to state that the phrase 'ecological services and amenities' would not be recognized by the majority of staff or management at WYM, with the possible exception of Jack Radford. 'Sustainability' and 'environmentalism' would be recognized, but an ecosystems perspective is not recognized.

However, there is an implicit recognition of the need to redress the ecological 'cost' of WYM's manufacturing approach. As a point of strategy and principle, WYM aims to progressively reduce its ecological footprint. In 2008, the company installed a wood pellet burner – a renewable form of energy – to take care of the water heating requirement during felting and the hot air used during drying the yarn. The water is heated to 90°C and the switch from electricity to wood pellet for the heating resulted in an annual saving of 50 000 kilowatt hours and an on-going reduction of CO_2 emissions of 350 tonnes per year. WYM has applied this innovative, renewable energy project in a commercial-scale production. This led to a renewable energy achievement award in 2009.

Another project related to ecological services was the 'flow down trial' that aimed to reduce the amount of water used to clean the yarn during felting. It involved a number of trials and different product types and the outcome was positive. The amount of water discharged as effluent was reduced by 30% and as a result there was a considerable reduction of chemicals used to clean the yarns, reducing the ecological impact of the business.

All these initiatives demonstrate the company's commitment to carbon reduction and it can only be an advantage, allowing WYM to harness these opportunities before many of its competitors in the markets.

Natural capital

In our LCA we included all of the processes that occur in the transformation of the raw materials (wool bales) into the finished felted carpet and rug yarn product. We also included the fuel use of employees travelling by car or flying to Europe, USA and Australasia for business purposes. We included CO_2 emissions of the various consumables used (cleaning, felting detergents, boiler chemicals and dyes) and of the packaging materials (packaged hanks and cones). Energy inputs and outputs (e.g. electricity consumed in the plant by the specialized production machinery as well as all the office equipment), and other forms of energy (e.g. wood pellets that provide the energy for heating water and drying wool) were also included. The transportation of the packaged felted yarn to the port was included.

We excluded the energy and emissions required to produce the raw material (wool bales) that arrives at WYM from wool suppliers. We also excluded the production and disposal of capital equipment (infrastructure, building and machinery); because it has a long service life, the environmental impact of its production and disposal is assumed to be very low. This is consistent with other studies (Ceuterick and Spirinckx, 1997; Barber and Pellow, 2006).

Wool waste was excluded, although it is recycled in other forms of products used in building, horticulture, animal health, transport and by humans. Also excluded were the steel straps that contain the wool bales. These are shredded and collected for recycling.

Wool bale wrapping materials together with the wooden pallets used for transport are returned to the wool suppliers and recycled. The impact of maintenance of production is assumed to be minimal and was excluded from this LCA. Finally, although we acknowledge that some dying takes place at the factory, the quantity of wool dyed is minimal (no more than 3% of the total quantity of wool produced) and therefore it is assumed here that the contribution to emissions is negligible. Odour and noise pollution, although acknowledged, were excluded.

The LCA evaluated WYM's carbon footprint between 1 September 2008 and 30 October 2009. The total CO_2 emitted was 44 965 kg. The firm produced a total of 269 tonnes of felted yarn during the studied period. Hence for the production of 1 tonne of yarn the CO_2 emissions were calculated to be 167.15 kg.

This figure is significantly lower than the reported 471 kg of CO_2 per tonne in previous work by Barber and Pellow (2006, p. 51) for the CO_2 emissions recorded from processing a tonne of wool top. Barber and Pellow's (2006) figure also included the cleaning of the wool through scouring (using hot water and detergent to remove grease and other impurities). At WYM the wool bales have already been scoured before they arrive at the factory hence it was not included within the system boundary of this study.

The figure of 167.15 kg of CO_2 per tonne of felted yarn was the cumulative emissions from all electricity used, wood pellets, the various consumables, fuel emissions and packaging during the period studied. To allow comparison with previous studies, the emissions from the carding process alone were estimated, which included the electricity for opening, blending, carding, merino combing and the consumable usage. Hence the total carding emissions produced for the period studied were calculated to be 13 067.76 kg of CO_2. As already mentioned, the firm produced a total of 269 tonnes of felted yarn during the studied period. Hence for the production of 1 tonne of felted yarn the carding CO_2 emissions were calculated to be 48.57 kg. The remaining 118.58 kg of CO_2 emissions were produced from the rest of the processes.

Total energy use over the period was 82 451 kWh in opening, blending and carding and a further 44 237 kWh for all remaining processes. In isolation this has limited meaning, but when converted to kg of CO_2 per tonne of felted yarn, it is apparent that what might be considered the 'preprocessing', which is comprised of opening, blending and carding (preparing the wool for the actual spinning processes), has, at 48.57 kg of CO_2 per tonne of felted yarn, easily the greatest ecological impact of all of the subprocesses that comprise the production of yarn.

Carding uses a chemical to lubricate the process. This contributes 0.3114 kg of CO_2 per tonne of felted yarn.

As we have earlier indicated, the hot water used in felting and the hot air used in drying the yarn is heated through a wood pellet burner. Wood pellets are a type of fuel that is made from compact sawdust. It is usually a by-product of saw mills and other similar wood transformation activities. The pellets are fed into a pellet burner and can be burned with very high combustion efficiency. Wood pellets are considered carbon neutral (the same amount of CO_2 emitted when burning pellets is consumed while a tree grows in the forest) therefore the use of wood pellets in the WYM has the advantage of considerably lowering the footprint by using them as heating fuel instead of the more traditional fuel oil. WYM's carbon footprint of wood pellets for 1 kg of felted yarn was on average 0.00935 g CO_2.

As we have indicated previously, water is a significant production factor at WYM. The local council bills WYM quarterly for charges relating to the waste water discharge from carding and felting. For the period 1 October 2008 to 31 December 2008 the company discharged 899.80 m³. Approximately 2000 L is used for the carding process daily, which yields approximately 1076 kg of felted yarn per day. The other process that is heavy on water usage is the felting process, consuming approximately 8000 L for producing 1076 kg of wool a day. Again as stated earlier, WYM has made some inroads in reducing their water discharge but perhaps this is an area for improvement.

The other components of the felting process are two chemicals: one a detergent, the other a dispersant (which disperses the detergent from the yarn). In the period of the LCA, the detergent contributed 0.196 kg of CO_2 per tonne of felted yarn, and the dispersant contributed 1.056 kg of CO_2 per tonne of felted yarn.

Transport is an essential element of WYM's operations in terms of the transport of goods and raw materials and in business travel, both domestic and international. In the period under study, raw materials and finished goods transport was excluded. Business travel by car was found to emit 5274 kg of CO_2 equivalent in the period under study. Domestic air travel accounted for 5340 kg of CO_2 equivalent and international air travel accounted for 9120 kg of CO_2 equivalent.

Human capital

The focus of the company on the production of premium goods is reflected in their investment in human and social capital. As part of this process, in March 2008, the board of directors appointed a dedicated general manager to manage the growth of all business operations and the future direction of the company. Mick, the general manager, brought with him 10 years of previous experience at senior executive level and a background in primary industry production processing and operations management.

The second in command is the productions manager, Mark, who has been with the company almost since its inception in 1995. He started as a machine operator and soon his knowledge and enthusiasm saw him move up to senior operator, team leader and his current role as a production manager running the factory and dealing with all production issues. During his time at WYM, Mark completed a national certificate in textile manufacturing and a national certificate in first-line management.

The product development expert and co-owner Ed is the brains behind the felting technology and the man who holds most of the technical knowledge and expertise on developing niche felted yarn products. With a background in textile science and technology, prior to forming WYM, he held a number of roles with the New Zealand wool board, including that of a technical manager. Ed thrives on new product development and he has a close relationship with all his customers, innovating not just in advance of their requirements but also in response to them.

Most of the 35 staff has extensive experience of the textile industry and many have enjoyed service with the company for over 5 years. The majority of the staff have been with the company for a long time and come from a textile processing background. Staff numbers have increased as sales have increased. The general approach of WYM towards staff is training to use best business practices, focus around health and safety, and improved quality and product knowledge.

To ensure best practices in product development, information is visually displayed in designated business areas. Each time an existing or new product is going through the factory, the monitoring visual board with all the important information on 'hints and tips' and 'how to run' the specific product is visually displayed for staff to familiarize themselves and achieve top-quality product each time, every time.

The objectives for Health and Safety, Engineering and Quality, has been set with the focus on having clear well-documented expectations that conform to benchmark standards, for example AS/NZS 4801, ISO 9000 or to internal standards. To that end, WYM is actively working to enhance their staff training programme by having written procedures for all process steps in felted yarn processing.

WYM has an integrity statement that sets behavioural expectations from staff towards the company's strategic goals and market position. Its integrity statement as stated in the *Operations Manual* is:

> the focus of all work carried out is to treat everyone with respect (customers, fellow employees, suppliers and other business stakeholders). This is achieved by accepting responsibility for tasks performed, ensuring the various processes are in control and meet product and best practices standards.

This statement is a prime example of the ethos and values that underpin the WYM culture, which is evident in the products, product support and market relationships. New staff complete an induction programme intended to introduce new employees to the company culture, and the relevant policies and procedures involved in their specific role as well as the company as a whole. The induction also includes appropriate tours to show local area health and safety measures, process areas, building exits and employee facilities. Managers and supervisors ensure that new employees are shown around their new work area appropriately, in addition to their induction training.

WYM actively encourages employee participation in company meetings to contribute to business productivity and effectiveness. During the company's production, and quality meetings staff are encouraged to raise any concerns, contribute their thoughts and overall add value to their products. Developing and maintaining staff commitment and enthusiasm is paramount in the business. Competent staff (especially senior operators or technical specialists) at the various production stages apply 'know-how' and record trial information including photos of best practices at each processing stage as appropriate.

The human resource strategy of the company is employee focused: nurture and build self-confidence in staff, provide support through processes and company personnel, notice efforts and reward them, and encourage participation in continual learning and improvement. Currently, WYM has appropriate core staff capable of moving up the ladder to more senior positions, although currently they are rolling out a training programme to enable all staff to clearly understand the baseline expectations.

The management philosophy is to empower staff to manage their own workspace successfully. To that end, Mick initiated the Competitive Manufacturing initiative, which lasted for 6 months and trained staff in principles of a clean and safe work environment. Among other things the course lead people to become smarter and work more productively in their area of expertise and improved communication among staff members.

Social capital

WYM's vision statement clearly indicates the need to:

establish long-term relationships with both suppliers and customers

WYM maintains a number of suppliers that it is dealing with on a regular basis. To qualify as a WYM supplier there is an initial screening of candidates, followed by a more thorough qualification process. Aspects to be covered with a supplier include: supply plan, remedy plan, payment plan, quality, production expectation and reference standards, unauthorized deviation/non-complying product and granting of concessions. A rating-based performance qualifies a supplier for delivering key items or processes.

WYM is a customer-driven organization. As Mick (General Manager) explains:

'Before we set any policy, before we spend any money, before we make any decisions, before we take any action, before we even open our mouth we should ask ourselves: If I was the customer, what would I want to see happen? Then that is what we should do.'

One of the first things that Mick tried to improve was the delivery on time. He introduced quality systems that would capture issues as they arise, record them and fix them. The Managed Event quality system at WYM has a 'Managed Event register book' for each defined process step, that is Card, Gill, Rove, Felt, Wind and Hank. All managed events are regularly checked by the team leaders and production manager, and records are reviewed with the emphasis on finding appropriate long-term preventative actions, especially if there are any reoccurring patterns to events. If there is a persistence or a pattern developing, then an Opportunity For Improvement (OFI) record would be created to capture the pattern, find solution, apply it and monitor outcomes. This great effort resulted in improving the delivery on time to reach 100% within a few months of Mick's appointment.

In addition to improving the delivery on time indicator, WYM also sought to establish a strong relationship with the customer, and become part of their product research and development team. With fostering strong relationships with customers, many of whom have been customers for many years, the company makes a conscious effort to improve the customer experience. WYM seeks feedback from customers on new product lines, and never pushes new products to customers; instead WYM's product expert generates innovative products from customer demands. Annual visits to customers from Ed and Mick are common to establish a good relationship and regular feedback. Through such attention to customers' needs, WYM continually attempts to enhance its relationships with them. As Mick says:

> 'We want to be at the front edge of it and by doing that, the customers and people we work with get access to good product development that keeps them at the front edge of the markets too.'

The company has restricted its degree of channel integration to include all processes between carding and yarn formation. A backward integration took place when WYM acquired a carding machine to guarantee continuous supply of carded sliver. Forward integration occurs when they acquire carpet or rug makers to assist in sales and distribution.

As demand for felted yarn products increased towards the end of 2010, the company strategic direction saw an invited merger with a major carpet manufacturer. The merger gave the manufacturer access to 75% of WYM's resources (technical and staff) while the remaining 25% of WYM services remaining clients.

Manufactured capital

For reasons of confidentiality we are not able to disclose detailed information on the capital base or wider financial information of WYM. However, it continues to heavily invest in technology and is at the forefront of developing new manufacturing equipment in felting and drying. Outside of these processes, WYM employs conventional spinning technology, some of which is imported directly by them, but other of which has been acquired from New Zealand spinners as their mills have closed.

WYM is a capital-intensive operation and this was one of the factors in the friendly takeover that took place in the course of our field study.

Community and individual well-being

In our fieldwork, other than what might be called 'normal' staff social functions (e.g. a Christmas party), we found no substantial empirical evidence to support any commitment by WYM to community and individual well-being or to the development of social capital, other than that required in law (e.g. occupational health and safety regulations). However, by observation of two of the present authors (Mary and Clive) both Jack and Ed have an observably paternalistic approach to their employees. For example, prior to the field study we report here, in the course of one of several recent recessions (the effects of which were exacerbated by huge exchange rate risk), Jack and Ed were forced to make several skilled staff redundant. The personal affront this caused Jack was readily evident in several conversations he and Clive had together at that time. Indeed as the recession turned out and trading conditions eased, most of the staff were re-employed.

Discussion

The case of WYM demonstrates that, through economic production, facilitated by a supply chain, natural capital may be transformed into goods and services, and wastes. Community and individual well-being outcomes though are less evident. Positive returns on natural capital are achieved when it is restored or conserved, but in this case, not through economic regulation, but through the twin desires to be more efficient and reduce ecological impact. Ecological and ecosystems services are not readily used concepts or terms at WYM. Neither community nor individual well-being is mentioned explicitly in our evidence. Waste is well managed and limited. However, in this case as in so many others, we have observed the sustainable sourcing of raw materials, inbound and outbound logistics, and efficient and effective manufacture remains largely within the boundaries of the firm. There was some evidence of a willingness to think beyond these constraints.

Facilitated by SCM through economic production, WYM's considerable human and social capital is transformed into goods and wastes. Positive returns to human capital are made through investment in learning and research, but that is mediated again through the twin desires to be more efficient and reduce ecological impact rather than by economic regulation. A limited amount of investment is lost to production waste, but much is conserved and much more social capital is generated in WYM's waste management partnership in waste recycling.

WYM offers evidence of highly skilled and committed individuals interacting in a highly professional manner to assure the efficient and effective manufacture of goods. WYM's committed investment in individual knowledge and skills as well as social networks exemplifies the commitment necessary to develop sustainable supply chains.

In WYM, SSCM is the common thread that ties ecological, social and economic performance elements together. This partially unintentional holistic approach to managing the economic production process, partially and only implicitly operationalizes the ecological economic model with its goal of sustainable human well-being.

Conclusion

We started out by asking: how can we move organizations to recognize and account for the provision of ecosystems services in the production of goods and services?

The answer, it seems to us, is a mixture of engaged education and research. That is, much of the sustainability education and research we have seen thus far, whilst of value, has failed to engage with manufacturers like WYM. Gaps in theory and practice eventuate as a result, because knowledge produced by researchers seems not to be relevant to practitioners. What is needed is participative research in sustainability and supply chain management to obtain the different perspectives of researchers, practitioners and stakeholders in studying this most complex and fundamental of all commercial practices. This involvement and the concomitant leveraging of different bodies of knowledge will produce more penetrating and relevant insights than when scholars work in isolation of practitioners (Van de Ven, 2007).

The answer then is that organizations will move to recognize and account for the provision of ecosystems services in the production of goods and services when we, as scholars, work with them. The work required is to develop grounded explanations of the ecological and social impact of supply chains, whilst recognizing a fundamental principle of business: to stay in business!

References

Barber, A. and Pellow, G. (2006). *Life Cycle Assessment: New Zealand Merino Industry Merino Wool Total Energy Use and Carbon Dioxide Emissions*. Available at: http://www.agrilink.co.nz/Portals/Agrilink/Files/LCA_NZ_Merino_Wool.pdf (accessed March 2012).

Beamon, B.M. (1999). Designing the green supply chain. *Logistics Information Management*, 12, 332–342.

Beamon, B.M. (2008). Sustainability and the future of supply chain management. *Operations and Supply Chain Management*, 1, 4–18.

Bowen, F.E., Cousins, P.D., Lamming, R.C. and Faruk, A.C. (2001). The role of supply management capabilities in green supply. *Production and Operations Management*, 10, 174–189.

Caldwell, N. and Smallman, C. (1996). Greening the value chain: operational issues faced by environmental management in the UK car manufacturing industry. *Eco-Management and Auditing*, 3, 82–90.

Carmody, J. and Trusty, W. (2005). Life cycle assessment tools. *Implications*, 5, 1–5. Available at: http://www.informedesign.umn.edu/_news/mar_v05r-p.pdf (accessed March 2012).

Carter, C.R. and Rogers, D.S. (2008). A framework of sustainable supply chain management: moving toward new theory. *International Journal of Physical Distribution and Logistics Management*, 38, 360–387.

Ceuterick, D. and Spirinckx, C. (1997). *Comparative LCA of Biodiesel and Fossil Diesel Fuel*. Vito, Mol, Belgium.

Costanza, R. (2000). Social goals and the valuation of ecosystem services. *Ecosystems*, 3, 4–10.

Curran, M.A. (2008). Life-cycle assessment. In: *Encyclopedia of Ecology* (eds J. Sven Erik and F. Brian), pp. 2168–2174. Academic Press, Oxford.

Drumwright, M.E. (1994). Socially responsible organizational buying: environmental concern as a non economic buying criterion. *Journal of Marketing*, 58, 1–19.

Finnveden, G. and Ekvall, T. (1998). Life-cycle assessment as a decision-support tool – the case of recycling versus incineration of paper. *Resources, Conservation and Recycling*, **24**, 235–256.

Finnveden, G., Hauschild, M.Z., Ekvall, T., et al. (2009). Recent developments in life cycle assessment. *Journal of Environmental Management*, **91**, 1–21.

Handfield, R., Sroufe, R. and Walton, S. (2005). Integrating environmental management and supply chain strategies. *Business Strategy and the Environment*, **14**, 1–19.

Hartmann, T. (2004). *The Last Days of Ancient Sunlight*. Three Rivers Press, New York.

Jeswani, H.K., Azapagic, A., Schepelmann, P. and Ritthoff, M. (2010). Options for broadening and deepening the LCA approaches. *Journal of Cleaner Production*, **18**, 120–127.

Lee, S.Y. (2008). Drivers for the participation of small and medium-sized suppliers in green supply chain initiatives. *Supply Chain Management*, **13**, 185–198.

Miles, M.B. and Huberman, A.M. (1994). *Qualitative Data Analysis. An Expanded Sourcebook*, 2nd edn. Sage Publications, London.

Min, H. and Galle, W.P. (2001). Green purchasing practices of US firms. *International Journal of Operations and Production Management*, **21**, 1222–1238.

Porter, M.E. (1985). *Competitive Advantage Creating and Sustaining Superior Performance*. Free Press, New York.

Rao, P. and Holt, D. (2005). Do green supply chains lead to competitiveness and economic performance? *International Journal of Operations and Production Management*, **25**, 898–916.

Seuring, S. and Müller, M. (2008). From a literature review to a conceptual framework for sustainable supply chain management. *Journal of Cleaner Production*, **16**, 1699–1710.

Simchi-Levi, D., Simchi-Levi, E. and Watson, M. (2004). Tactical planning for reinventing the supply chain. In: *The Practice of Supply Chain Management: Where Theory and Application Converge* (eds T.P. Harrison, H.L. Lee and J.J. Neale), pp. 14–30. Springer, New York.

Srivastava, S.K. (2007). Green supply-chain management: A state-of-the-art literature review. *International Journal of Management Reviews*, **9**, 53–80.

Staib, R. (2009). *Business Management and Environmental Stewardship*. Palgrave Macmillan, London.

Strasser, K.A. (2011). *Myths and Realities of Business. Environmentalism: Good Works, Good Business or Greenwash?* Edward Elgar Publishing, London.

Svensson, G. (2007). Aspects of sustainable supply chain management (SSCM): conceptual framework and empirical example. *Supply Chain Management*, **12**, 262–266.

Vachon, S. and Klassen, R.D. (2006). Extending green practices across the supply chain. *International Journal of Operations and Production Management*, **26**, 795–821.

Van de Ven, A.H. (2007). *Engaged Scholarship. A guide for Organisational and Social Research*. Oxford University Press, Oxford.

van Hoek, R.I. (1999). From reversed logistics to green supply chains. *Supply Chain Management*, **4**, 129–135.

Zhu, Q., Sarkis, J. and Geng, Y. (2005). Green supply chain management in China: pressures, practices and performance. *International Journal of Operations and Production Management*, **25**, 449–468.

11

Market-based Instruments and Ecosystem Services: Opportunity and Experience to Date

Stuart M. Whitten[1] and Anthea Coggan[2]

[1] CSIRO Ecosystem Sciences, Canberra, Australia
[2] CSIRO Ecosystem Sciences, Brisbane, Australia

Abstract

Market-based instruments (MBIs) are measures designed to signal to land managers the rewards or consequences for certain actions in much the same way as markets do. MBIs offer a new way for governments to bring about an increased level of supply of environmental goods. World-wide attention to MBIs is increasing because of their potential to deliver environmental outcomes at lower cost to government and market participants (cost effectiveness), allow flexibility in individual response (are efficient), and encourage positive environmental outcomes (positive rather than negative incentives). Together these advantages can drive innovation and dynamic efficiency in delivering desired ecosystem service outcomes.

In this chapter we discuss the different MBI opportunities available and provide examples of each. We note that MBIs are not the panacea for all environmental problems. Their performance relative to other approaches is reliant on appropriate selection, design according to the outcomes sought, the nature of the market failures faced, and the natural and human environment in which the policy will operate. Furthermore, all policy interventions are costly, in terms of design, delivery, impacts and any incentives provided. These costs should always be compared against the alternative of doing nothing.

Ecosystem Services in Agricultural and Urban Landscapes, First Edition. Edited by Steve Wratten, Harpinder Sandhu, Ross Cullen and Robert Costanza.
© 2013 John Wiley & Sons, Ltd. Published 2013 by John Wiley & Sons, Ltd.

Introduction

Many countries around the world have been undergoing a long period of land-use change and development motivated by the significant private values received from activities such as land development, grazing, agriculture and mining. This land-use change has degraded many of our environmental and cultural assets and the ecosystem services that they produce (Sala et al., 2000; Srinivasan et al., 2008). Despite their obvious importance to our ongoing well-being, ecosystem services have largely been ignored in both domestic and international markets, law and policy (Ribaudo et al., 2010). Instead, existing markets have rewarded agriculturalists, miners and others but 'failed' to conserve environmental and cultural goods. That is, most markets do not transmit clear signals that encourage participants to use and manage natural resources sustainably, most often because of factors such as incompletely described or enforced rights and entitlements, high transaction costs or other market impediments. As a result, the full costs of production decisions are not reflected in the market price paid where ecosystem services are concerned. A simple example is producing a tonne of wheat. The price paid for the wheat does not include any costs of lost environmental services due to land conversion (from forest to crop land) or ongoing externalities such as environmental degradation generated from inappropriate farming practices.

In theory, the supply problems for goods arising from market failure can be remedied through some level of government intervention (Murtough et al., 2002). Intervention can be divided into three distinct categories:

- **Facilitative:** where measures are designed to improve the flow of information and corresponding signals and incentives without providing any direct incentive payments to landowners. For example, extension programmes providing information about how to manage land to improve biodiversity conservation.
- **Incentive:** where measures are designed to directly alter the structure of pay-offs to land managers and are usually specifically intended to substitute for missing monetary signals that are generated within markets for other goods and services. Pollution taxes, environmental subsidies and various forms of payment schemes are examples.
- **Coercive:** where non-voluntary measures are designed to compel management change using the coercive powers of government. Regulations designed to protect native vegetation or threatened species are an example of coercive policies.

Market-based instruments (MBIs), which primarily fall into the 'incentive' category of intervention,[1] are just one way for government to bring about an increased level of supply of environmental goods. MBIs are receiving increasing attention world-wide because they have the potential to deliver environmental outcomes at lower cost to government and market participants (cost effectiveness), allow flexibility in individual response (are efficient), and encourage positive environmental outcomes (positive

---

[1]Some MBI forms require coercive elements to structure the market and create demand. For example cap and trade or offset type schemes.

rather than negative incentives) (Stavins, 1998; Freeman and Kolstad, 2007; Whitten et al., 2007; Ribaudo et al., 2010). Together these advantages can drive innovation and dynamic efficiency in delivering desired ecosystem service outcomes.

That said, MBIs are not the panacea for all environmental problems. Any policy approach intended to achieve a better supply of environmental and cultural goods needs to be carefully designed according to the outcomes sought, the nature of the market failures faced, and the natural and human environment in which the policy will operate. Furthermore, all policy interventions are costly, in terms of design, delivery, impacts and any incentives provided. These costs should always be compared against the alternative of doing nothing. That is, in many cases the costs of acting to remedy environmental degradation may be greater than the costs of the degradation to the community.

The remainder of this chapter is set out as follows. We first define what MBIs are and identify generic conditions under which they are likely to be successful. Next we describe the range of MBIs available and their conceptual structure. A range of example markets for ecosystem services are then described before the chapter closes with some observations on key design requirements and future research required to support successful MBI design and delivery.

Market-based instruments: definition and preconditions

MBIs are policy tools that encourage certain behaviours through market signals rather than through explicit directives such as regulation (Stavins, 2000). Well-designed MBIs target the cause of market (and government) failures. For environmental goods this is primarily the lack of fully defined and enforceable property rights compounded by high transaction costs and other market impediments. MBIs are intended to alter the pay-offs faced by land managers for various land management decisions. In the same way regular markets tend to influence people's behaviour, MBIs use trading mechanisms to deliver price signals and ultimately to influence behaviour that will deliver ecosystem services. These services may include the conservation of biodiversity, carbon sequestration and improvements in water quality, amongst others.

Whilst MBIs have the potential to provide environmental outcomes more efficiently than other options such as command and control approaches, they do work best in certain circumstances. Some contextual questions to ask to gauge whether circumstances would support MBIs are provided in Table 11.1.

Types of MBIs

There are many ways to describe and categorize MBIs. Most typologies distinguish between price and quantity-based instruments. A third group of instruments aimed at improving the operation of existing markets, termed 'market-friction' instruments, are also sometimes included in MBI discussion (NMBIPP, 2004).

Table 11.1 Contextual questions to gauge if an MBI is an appropriate policy response.

Are there significant gains from trade available to drive a market?	A market mechanism can only function where there are potential 'gains from trade'. Gains from trade are primarily realized because of heterogeneity. Heterogeneity exists within the landscape (services may be located across the landscape or located in particular areas 'hotspots'), between different management actions (different landowners can undertake different actions to address the same NRM issue), and where social and economic variation exists between landholders (landholders experience different cost structures and have different preferences). Heterogeneity drives the gains from trade.
What is the extent of transaction costs of capturing the gains from trade?	Trades will only occur where the value of the relevant service outweighs the sum of production and transaction costs incurred in the market. Transaction costs are those costs associated with buying and selling, such as those associated with collecting information and processing trades. Thus MBIs are only a practical option where the service generates sufficient value to encourage trade and where transaction costs can be sufficiently minimized to facilitate market exchange.
What are the existing policies and schemes and their implications for MBI design and potential?	No policy instrument or reform is truly 'new' since it must be superimposed or partly replace existing rules, regulations and customs. Thus in crafting policy instruments, it is helpful to think of them as complementing or amending the *status quo*. One must consider not only the proposed policy instrument but the current institutions and operating frameworks and whether they need to or can change.
What is the community capacity to support the MBI?	Policy is not only generated within existing rules, regulations and customs but also within constraints and opportunities provided through existing political structures, biophysical constraints and physical, human, financial and social capitals. One must consider these contextual attributes in assessing whether the policy opportunity and the policy instrument can be adapted to achieve the desired outcome.

Price-based approaches set or modify the price of environmental impacts within existing markets through payments, charges, taxes or subsidies. Firms then respond to the modified market signals by adopting the resource use or management practice that offers them the greatest benefit and, if the policy is effective, improving ecosystem service outcomes. While these instruments cannot guarantee the *extent* of changes, they act to cap the *costs* incurred under the instrument.

Price-based instruments therefore rely primarily on price signals rather than scarcity to create incentives to potential participants. They are reliant on a source of funds (from which payments are made) or a legal basis on which to alter prices (via taxes or subsidies). Hence, most price-based instruments can be referred to as single-sided (single buyer) or no-sided instruments (uniform tax or subsidy) rather than a competitive two-sided market with many buyers and sellers (Whitten and Young, 2004). Single-sided instruments employ many sellers but only a single buyer, as is the case with procurement auctions for biodiversity. No-sided instruments involve a voluntary decision about whether to engage but a uniform price signal, as is the case with taxes and subsidies. International literature tends to consider some sort of competitive access, based for example on quantity of ecosystem service provided, to be a necessary condition to be considered an MBI (including single-sided markets). Non-competitive, open access type approaches are generally excluded from MBI frameworks (including most no-sided markets). We follow this convention in this chapter by defining price-based MBIs as requiring the purchaser to decide (or negotiate) what is to be purchased (and how it will be measured), agree a payment to be made, and how the resultant contract will be monitored and enforced.

Quantity-based or 'tradable rights' instruments create a market in the right or entitlement to engage in an activity associated with specified resource uses or environmental damage. They do this by restricting the total level of activity and allocating rights to participants. An efficient allocation of rights is then determined through market exchanges. Quantity-based MBIs usually involve many buyers and sellers (and so are considered to be double-sided markets).[2] Tradable rights instruments tend to be used when it is important to get a certain environmental outcome (for example, when pollution of a waterway is close to a threshold level that may cause irreversible or unacceptable degradation). Government or a designated authority must determine how to measure the ecosystem service or damaging activity, the total quantity of the goods to be expressed in the rights or entitlements, who can own them, the initial allocation, the conditions under which trade can take place, and how rights and entitlements will be monitored and enforced (Schwarze and Zapfel, 2000; Murtough et al., 2002).

Quantity-based MBIs tend to require legislative changes to create and enforce rights and entitlements and can persist so long as the institutional environment is maintained. In contrast, price-based MBIs can be implemented using existing contract law but require a source of funding and can only persist so long the funding is available. Therefore the institutional complexity associated with quantity-based MBIs tends to be higher than for price-based schemes.

Market friction mechanisms work to improve the way a current market functions. The intention is to reduce transaction costs and thus facilitate improved market signals. Market friction MBIs tend to focus on the key market impediments outside rights and entitlements or direct lack of a price signal. Typical foci are information provision (about the source of ecosystem services or impacts of activities on ecosystem services), facilitation of a market place or trading environment, or reducing impediments to participation in markets (including capital, knowledge and other constraints). An example of a market friction MBI is the creation of a water exchange or subsidization of brokerages to improve water market outcomes.

---

[2] There are exceptions which can be considered one-sided quantity-based markets. For example, an auction of pollution rights (for example carbon emission permits) is a one-sided market with one seller and many buyers.

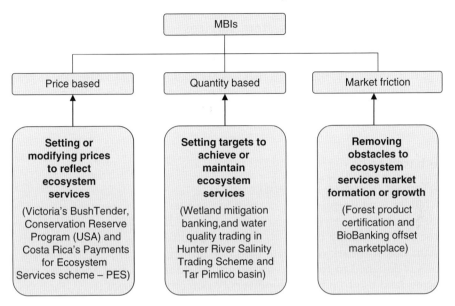

Fig. 11.1 Types of market-based instruments (MBIs). (Based on data from *Managing our Natural Resources: Can Markets Help? National Action Plan for Salinity and Water Quality – National Market Based Instruments Pilots Program*. Australian Federal Government.)

Table 11.2 Examples of MBI types.

Price based	Quantity based	Market friction
Emission charges	Cap and trade schemes	Revolving funds
User charges	(may employ permits,	Market/ contract
Stewardship payments	entitlements, quotas or	protocol development/
Payment for ecosystem	similar)	sponsorship
service schemes	Permit auctions (within	Insurance (risk) schemes
Performance bonds	and outside cap and	Research programmes to
Non-compliance fees	trade)	support market
Removal of perverse	Offset schemes	development
incentives		Labelling schemes
Deposit-refund schemes		Information disclosure/
		mandatory reporting

The MBI typology is illustrated in Fig. 11.1 with some examples of broad MBI types that may be applied to ecosystem services in Table 11.2. As the examples in Table 11.2 demonstrate, many MBI schemes may employ several elements across the typology. For example, a market friction oriented loan scheme supporting investment in ecosystem services for future sale could also be seen as a price-based mechanism. Similarly, permit auctions may be considered a price scheme because they place a price on the ecosystem service (or damaging action) as much as they are a quantity-based outcome (because damage is limited to the permits sold).

Examples of MBIs for ecosystem services

In this section we present a number of examples of successful MBIs as an indication of the range of market forms that have been developed to date. These are just a subset of the possible market forms and it is important that MBIs are designed and implemented to meet the specific market and government failures that are present in the particular environment for which an MBI is being considered.

Price-based MBIs

BushTender (Victoria, Australia)

BushTender is an example of a competitive tender or procurement auction MBI which purchases ecosystem management from landholders. The BushTender approach was initially piloted to improve biodiversity stewardship actions for remnant Box Ironbark on private land (see Stoneham et al., 2003). The auction process is designed to reveal asymmetrically held information about the costs of changing land use to improve biodiversity outcomes (known to landholders but not to government); and the biodiversity benefits from changing land management (known to government but not to landholders). The BushTender programme has subsequently been expanded across Victoria and is now the default option for investment in biodiversity on private land.

BushTender works by asking landholders to offer a price (a bid) to undertake part or all of a set of desired management actions. A biodiversity benefits index is used to measure the benefits delivered by the change of management. Bids are ranked by the biodiversity benefits per dollar and offered to landholders until the budget is exhausted (or until bids are no longer considered cost effective). Individual management contracts are used to specify required actions, payments and monitoring arrangements. Evaluation of the BushTender Box Ironbark pilot indicated that a fixed price scheme, would achieve 25% less biodiversity than the competitive tender if operated with the same budget (Stoneham et al., 2003).

Conservation Reserve Program (CRP) (USA)

The CRP, an example of an incentive payment scheme, was created in 1985 to fulfil the dual purpose of reducing supply of crop land (bolstering agricultural prices) and reducing soil erosion. The CRP provides an annual rental payment to landholders who agree to retire land from agricultural production (Perrot-Maitre, 2001). Since 1996, the CRP has evolved to encompass a broader range of ecosystem services such as wildlife habitat, air and water quality objectives and enrolment criteria. Farmers wishing to enrol in the CRP have their offers ranked by government field officers according to an environmental benefits index (EBI). The EBI is a composite score, with points for (Cattaneo et al., 2006):

- wildlife habitat benefits resulting from vegetation cover on contract acreage;
- water quality benefits from reduced erosion, runoff and leaching;
- off-farm benefits of reduced erosion;
- benefits that will likely endure beyond the contract period;

- air quality benefits from reduced wind erosion;
- benefits of enrolment in conservation priority areas; and
- cost.

While the CRP has many supporters, there are also many criticisms of this scheme. First, there are concerns over leakage – farmers may be ploughing up other land to compensate for land placed in the CRP programme. For example, nine Great Plains states enrolled 17.3 million acres in the CRP from 1985 to 1992 but the total amount of harvested cropland only declined by 2.6 million acres (Wu, 2000). Salzman (2005) also provides some criticism of the CRP, noting that the land eligibility criteria may be interpreted too broadly, allowing CRP enrolment for lands that do not need to be set aside. For example, as much as 77% of the CRP land in Minnesota could be farmed with little ecological harm if proper management practices were used. Further, the programme can send the wrong message. Many farmers note that due to past stewardship they now do not qualify for CRP funds even though their land was intrinsically as erodible as their neighbour's. Finally, there is evidence that farmers have colluded in the bidding process, all setting bids just below the programme's maximum acceptance price (which was set as a per hectare price) and well above local market rental rates (Salzman, 2005). Similar criticisms potentially apply to most payment for ecosystem service schemes (PES), illustrating the requirement for care in design and delivery.

Payments for ecosystem services (PES) (Costa Rica)

Costa Rica's PES scheme, Pago por Servicios Ambientales (PSA) is an example of a price-based market with multiple actions, multiple outcomes and multiple buyers of these outcomes. Ecosystem services purchased include water quality, biodiversity, scenic amenity and greenhouse gas mitigation. The buyers include government and direct beneficiary purchasers who voluntarily act together to purchase ecosystem outcomes. For example, in 2008 there were 11 contracts for provision of water services (for water quality) under Costa Rica's PSA programme (Pagiola, 2008). Those that are paying, use water for hydropower production, drinking water (water bottler), municipal water supply, irrigation and recreation. In 2005, the voluntary user-pays water-quality scheme generated $US0.5 million annually.

In 2005, Costa Rica expanded the use of water payments by revising the water tariff paid by water users who hold water-use permits and introducing a compulsory conservation fee. This was expected to generate $US19 million annually with 25% of this to go to the PSA programme (including for other ecosystem services). Despite the increased revenue available, the compulsory fee and centralized government payment scheme is likely to lead to increased transaction costs due to increased costs of programme delivery and loss of privately supplied monitoring and enforcement services (Pagiola, 2008).

While water quality is the primary target of PSA, the scheme also purchases biodiversity and carbon services from landholders. Funds for biodiversity are sourced through the Global Environment Facility (GEF) and carbon services from compulsory fuel tax revenues. The risk of this approach is that payments may not be ongoing. Efforts have been made to generate a user pays process for biodiversity from the tourism industry but without success to date. Pagiola (2008) reports that the beneficiaries of landscape services tend to be fragmented and numerous, which makes it difficult to generate collective action in securing payments.

Quantity-based MBIs

Wetland mitigation banking (USA), BushBroker (Victoria, Australia)

Offsets typically involve the removal of threats via protection of high conservation value habitat or the restoration of degraded habitat to compensate for the loss of habitat in nearby locations. They effectively replace damage to ecosystem services by requiring equivalent beneficial restoration as compensation. Arguably one of the best known examples of an offset scheme is the wetland mitigation programme driven by the *Clean Water Act (1972)* in the United States. Under certain conditions a developer is allowed to substantially alter a wetland if they undertake to protect, restore or enhance an equivalent amount of wetland ecosystem services elsewhere (US Environment Protection Agency (EPA), 2012). The early operation of the scheme indicates that whilst in theory an offset should be a more efficient management option compared with regulation, poor design could result in negative environmental outcomes (see Coggan et al. (2010a) for more detail on design and the consequences for offset schemes). Initial wetland offsets have been particularly criticized for poor equivalence metrics – the specification of what can be traded for what. For example, a wetland offset scheme in Maryland, which only used wetland acres rather than wetland ecosystem services in offset trades, gained 122 acres of wetland area through mitigation from 1991 to 1996 but lost the wetland function equivalent to a loss of 51 acres (Salzman and Ruhl, 2000).

The BushBroker native vegetation credit registration and trading programme, established in 2006 by the Victorian state government in Australia, provides a differing example of offset approaches. The biodiversity offset approach emerged as a response to regulation of native vegetation clearing in 1989 which provided a 'cap' on damage (though there are substantive exemptions). BushBroker was designed to allow economic development while delivering on the Victorian government's 1997 Biodiversity Strategy objective of net gain in extent and quality of native vegetation. In contrast to wetland mitigation banking, an integrated quantity and quality oriented process to quantifying and trading vegetation offsets and credits was adopted. This process was facilitated via a comprehensive exchange metric which encompasses all vegetation types in Victoria and was developed based on the Habitat Hectares approach (Parkes et al., 2003). Credits are generated as a result of improved management or security for remnant vegetation patches or revegetation relative to a predetermined baseline. Over 100 BushBroker credit agreements had been completed by 2010 (DSE, 2010).

There are a number of broadly similar offset schemes underway or under consideration for species, habitat, stream flow and other ecosystem services across Australia and internationally (see for example Bauer et al., 2004; Fox and Nino-Murcia, 2005; Gibbons et al., 2009). A consistent criticism of offsets relates to service equivalence and metric design and illustrates the difficulty in appropriate measurement of ecosystem service provision. For example, McCarthy et al. (2004) support the use of quantitative metrics for offsets but critique the habitat hectares approach on a number of grounds such as the use of distance and vegetation type as well as the use of benchmarks. Zedler (1996) asks similar questions of wetland mitigation banking, concentrating on the difficulties to measure function in wetland offsetting projects.

Hunter River Salinity Trading Scheme (HRSTS)
(New South Wales (NSW), Australia)

The HRSTS is an example of a cap and trade scheme linked to a regulatory approach limiting damaging salinity pollution into the Hunter River, NSW Australia. The HRSTS is intended to avoid damaging impacts on water quality and environmental outcomes from point source salinity pollution. Water quality impacts resulted from a combination of reduced flows due to extraction for irrigation and electricity generation cooling, and elevated salinity levels due to discharge of saline water into the river, primarily from mining activities. Water extraction licences (which are tradable) are intended to manage impacts of water extraction while a cap and trade approach (the HRSTS) has been introduced to manage salinity from point sources. The HRSTS is the focus here.

The HRSTS divided the river into management unit blocks (known as reaches) with a specified maximum salinity level in each block. Polluters were initially allocated a proportional discharge right which could be exercised under specified flow and water quality conditions. When river flow is low (usually with elevated salt levels) discharge is not permitted. Alternatively, when river levels are high (usually with low salinity levels), saline water discharges are allowed subject to salt loads and permit levels.

Businesses wanting to discharge saline water into the river must be licensed. Originally these licences were issued to existing licensed discharge points. In part to ensure that new entrants are able to access the market, salinity credits are now time limited (5 years) and a proportion are auctioned annually. Credits can also be traded thereby encouraging credit users to reduce salinity impacts and sell unused credits (Australian Government, 2008; Olmstead, 2010).

Tar Pamlico River Basin non-point water quality trading
(North Carolina, USA)

Point to non-point source trading instruments set a cap on the pollution levels of identified and licensed point-source polluters (e.g. factories, municipal wastewater treatment plants) but allow these point-sources to contract with non-point-source polluters to make the reductions on their behalf. Contracting farmers agree to reduce nutrients entering streams in the watershed by introducing buffers and changing land management practices (Olmstead, 2010). The cap placed on the point-source polluters establishes the demand for pollution reduction. Point to non-point-source trading has increased in popularity due to the widening gap in abatement costs between the two polluter types (Freeman, 2000). In the United States, the EPA has estimated that point non-point-source trading could reduce the cost of compliance to the Total Maximum Daily Loads (TMDL) regulations by $US1 billion or more annually between 2000 and 2015 (US Environment Protection Agency (EPA), 2001).

One of the most successful point non-point trading programmes in the United States is located in the Tar Pamlico River Basin, North Carolina. Point sources purchase agricultural nutrient reduction credits through an intermediary (the Tar Pamlico Basin Authority).[3] In 1999, the credit price was $US29 per kilogram of nitrogen or phosphorous; by 2010, nitrogen had fallen to $20 per kilogram while phosphorus had increased to $312.[4] In 1999, the estimated reduction in nitrogen and phosphorous of the point-source polluters was $US55 to 65 per kilogram (Olmstead, 2010).

One potential issue with point non-point trading schemes is the level of uncertainty surrounding management actions and their ability to reduce pollution. To overcome this, regulators often require more than one unit of non-point-source reduction for each unit of credit required in a point-source emission reduction (Olmstead, 2010). Available point-source demand may also limit the potential for trades (Ribaudo and Nickerson, 2009).

Market friction MBIs

Ecolabelling

'Ecolabelling' and 'green marketing' are tools that differentiate between products by drawing attention to positive environmental performance such as contribution to the provision of ecosystem services. Ecolabelling for environmentally friendly management is designed to benefit producers through increased market share or gaining premium prices for their products. Ecolabelling is a form of MBI addressing the problem of information failure. This type of MBI has been applied to single products (Banrock Station Wines, Australia), commodities (timber) and regions (King Island products, Australia), usually in the form of product labels.

Certification of forest products is a common ecolabelling example, focusing on sustainable forest management. Rametsteiner and Simula (2003) indicate that certification provides a positive impact on sustainability and forest management and the longer-term potential for this approach, but also suggest caution given the limited overall impact of forest product certification to date, particularly in tropical regions where it is most needed. Rametsteiner and Simula (2003) and Auld et al. (2008) also describe a number of challenges facing ecolabelling including: difficulty in defining, describing and measuring the effectiveness of sustainability objectives; the plethora of competing labelling schemes driven by differing local and international agendas; the potential for labelling to be directed towards other uses such as trade barriers; and the difficulty of poor countries and smaller firms to achieve the institutional and other requirements for labelling. Overall, it seems unlikely that ecolabelling will succeed in protecting ecosystem services without other supporting drivers.

Market places

A common problem for emerging markets is a lack of identified process or location (physical or other) for trading. Designer MBI approaches often assist

---

[3] The use of a trading intermediary also demonstrates the use of a market friction tool, to reduce the transaction costs of trades, within the quantity-based MBI.
[4] North Carolina Ecosystem Enhancement Programme fees as at 1st September 2010 (http://www.nceep.net/pages/resources.htm).

in creating new market places by providing model processes and a dedicated market place facilitation strategy. Model processes often include specified eco-system service attributes (sometimes including certified assessment processes) and model contracts. Market places range from relatively simple bulletin boards associated with approval processes, to complex smart market type approaches (such as the BushBroker exchange). In new or emerging markets ecosystem service provision can benefit from fostering market place develop-ment in order to achieve market liquidity and stable trading patterns sooner than they would otherwise emerge. Examples of these types of market friction MBIs include:

- The US New Jersey Pinelands transferable development rights scheme where the government has taken on a feeless brokerage role (Stavins, 1995); and
- The NSW BioBanking offset scheme and the Victorian BushBroker offset scheme operate with the government agency providing an offset seller/ pro-vider matching service.

In many instances independent brokers quickly emerge to aid in the exchange process via knowledge or other benefits they have over first time or infrequent participants (Coggan et al., 2010b). These brokers may also create and manage market places, as was the case in sulphur dioxide trading market (Stavins, 1998) and also are common in water markets in south east Australia (see for example Chapter 3 of Productivity Commission, 2010).

The brave new world of ecosystem markets

Designing effective MBIs

Despite the attention that MBIs are receiving around the world as a promising mechanism to encourage landholders to supply ecosystem services, their effec-tiveness is strongly influenced by application context and design. As our experi-ence in their design and application grows, some key features for their successful design and implementation are emerging. These include: recognizing and har-nessing the gains from trade; knowing where to start the design process; a focus on addressing the impediments to market formation in the design process; cost-effective measurement of service provision; and incorporating appropriate supporting measures.

MBI benefits result from harnessing the 'gains from trade'. Gains are derived from differences, or heterogeneities, amongst landholder preferences, resources or production opportunities. Future gains are captured by creating positive incentives to improve management rather than to avoid regulation, and by encouraging innovation. Where these gains cannot be harnessed an MBI will perform no better, and may perform worse than other measures. Good informa-tion about the characteristics of the desired ecosystem service and the potential producers underpins assessment of the potential gains from trade at all stages of design. At an early stage it helps to describe market boundaries by describing the degree to which different actions or actions at physically separate locations can be substituted in order to achieve the desired outcome.

Decisions about which MBI form is appropriate are initially based on whether existing markets are present and can be modified compared to creation of entirely new initiatives. Decisions between price and quantity-based MBIs are based on economic and non-economic factors. Economic factors include: the relative marginal costs and benefits; thresholds in cost or benefit functions; and number of market participants (Weitzman, 1974; Whitten et al., 2007). Non-economic factors include: property right preferences (duty of care, polluter pays or beneficiary pays); time lags in production; jurisdictional powers; and transaction costs of the instrument (Whitten et al., 2007).

MBIs are intended to overcome market and government failures and other impediments to market formation in order to release the gains from trade. The range of such failures present should be systematically identified as a basic input into the MBI design process (see for example Whitten et al., 2007). Incomplete property rights and information failure or asymmetry are likely to be present in all cases. Core property right and information asymmetry issues tend to be compounded by other market failures and design issues. Design contexts vary widely and so there may also be other, potentially unique, factors to be overcome.

Being able to measure what is being traded is essential to the function of the MBI. The role of the measurement metric in an MBI is often confused because of the multiple roles that measurements of environmental assets, ecosystem services and management actions play in the natural resource management (NRM) sphere. Construction (or adoption) of an appropriate metric is a critical requirement for measuring relative and absolute outcomes, and consequently who benefits and who pays.

Finally, MBI design must incorporate the necessary supporting mechanisms needed to ensure success, such as regulatory change, or communication and engagement programmes. Opportunities to nest MBIs within existing institutional and organizational architectures in order to reduce transaction costs should also be identified where appropriate.

Where to next in the brave new world of markets for ecosystem services?

Price, quantity and market friction style MBIs are relatively new to the toolkits of government and others in encouraging private production of ecosystem services. These markets are growing in popularity due to their potential to deliver environmental outcomes at lower cost, enhanced flexibility, incentive compatibility and prospective innovation and dynamic efficiency in delivering desired ecosystem service outcomes.

Despite the progress to date it is clear that substantial obstacles remain to the widespread adoption and growth of ecosystem service markets. These obstacles are presented by a combination of the characteristics of ecosystem service commodities, the characteristics of existing institutional frameworks, and incomplete information. A key characteristic of many ecosystem services are their public good attributes; non-excludability and non-rivalry.

The institutional environment is an important attribute of excludability and exploration of novel right and entitlement structures offers potential for market development. New rights, entitlements and obligations can be created and new technologies for measuring and tagging ecosystem services production and consumption invented as illustrated by the proliferation of cap and trade type approaches. Nevertheless, it is not always easy to link service provision to individual entities without which it is difficult to create effective markets. Similarly, non-rival consumption can be overcome where individual willingness to pay can be captured or exceeds cost of production. However, where there are large numbers of joint consumers, creation of an effective market remains likely to be more costly (in terms of transaction costs at least). In both areas a key opportunity lies in cost-effective measurement of ecosystem service provision.

An overlapping complication with respect to many ecosystem services is the inconsistency between existing institutional boundaries and those necessary for ecosystem service provision (and in some cases consumption). There may be scale thresholds in ecosystem service production which do not align with existing property or management boundaries. For example, effective biodiversity conservation may require coordinated action across property, jurisdictions and even internationally (such as for migratory species). Similarly, there may be jurisdictional or international borders between service providers and consumers complicating the potential for a designer market response. Solving boundary issues is likely always to be difficult but there are immediate opportunities for designing market rules that appropriately interact with biophysical and ecological processes. As an example, fishery stock growth rates (reproduction rates) are critical parameters in setting appropriate individual tradable quotas in fisheries which are linked to population dynamics.

The rapid emergence and growth of MBI approaches signals the role that knowledge about markets plays in identification, design, implementation and evaluation of new markets. Nevertheless, there remains much to be learnt about the ways in which markets operate and about how individual entities and markets interact. We are certain to devise new market forms, technologies to lower transaction costs in markets, and opportunities to more effectively monitor and enforce rights and entitlements. Investment in understanding markets, individuals and their interaction will influence the development and form of markets for ecosystem services in unexpected ways in the future.

Finally, but most importantly, the science uncertainties in ecosystem service production are important influences on the potential for markets. If we do not know where and how ecosystem services are produced it is difficult to design appropriate mechanisms, market based or other, to encourage protection and production.

References

Auld, G., Gulbrandsen, L.H. and McDermott, C.L. (2008). Certification schemes and the impacts on forests and forestry. *Annual Review of Environment and Resources*, **33**, 187–211.

Australian Government (2008). *Designer Carrots Factsheet: Cap and Trade Mechanisms.* Available at: http://www.marketbasedinstruments.gov.au (accessed August 2012).

Bauer, M., Fox, J. and Bean, M. (2004). Landowners Bank on conservation: the US fish and wildlife services guidance on conservation banking. *Environmental Law Reporter*, 34, ELR 8-2004.

Cattaneo, A., Hellerstein, D., Nickerson, C. and Myers, C. (2006). *Balancing the Multiple Objectives of Conservation Programs*. Economic Research Service, USDA, Washington DC.

Coggan, A., Buitelaar, E. and Whitten, S.M. (2010a). *Transferable Mitigation of Development Impact: The Case of Development Offsets at Mission Beach, Australia*. Conference paper presented at the International Academic Association on Planning, Law, and Property Rights in Dortmund (Germany). Available at: http://www.csiro.au/people/Anthea.Coggan.html (accessed August 2012).

Coggan, A., Buitelaar, E. and Whitten, S.M. (2010b). *Third Parties in Development Offset Markets: What Brings Them In?* Paper presented at the 4th World Congress of Environmental and Resource Economists (WCERE), Montreal, June 2010.

DSE (2010). *BushBroker*. Department of Sustainability and Environment, Victoria. Available at: www.dse.vic.gov.au (accessed August 2012).

Fox, J. and Nino-Murcia, A. (2005). Status of species conservation banking in the United States. *Conservation Biology*, 19, 996–1007.

Freeman, A. (2000). Water pollution policy. In: *Public Policies for Environmental Protection* (eds P. Portney and R. Stavins), 2nd edn, pp. 169–213. Resources for the Future, Washington, DC.

Freeman, J. and Kolstad, C.D. (2007). Prescriptive environmental regulations versus market-based incentives. In: *Moving to Markets in Environmental Regulation: Lessons from Twenty Years of Experience* (eds J. Freeman and C.D. Kolstad), pp. 3–18. Oxford University Press, New York, NY.

Gibbons, P., Ayers, D., Seddon, J., et al. (2009). An operational method to rapidly assess impacts of land clearing on terrestrial biodiversity. *Ecological Indicators*, 9, 26–40.

McCarthy M. A., Parris K. M., van der Ree, R., et al. (2004). The habitat hectares approach to vegetation assessment: an evaluation and suggestions for improvement. *Ecological Management and Restoration*, 5, 24–27.

Murtough, G., Arentino, B. and Matysek, A. (2002). *Creating Markets for Ecosystem Services*. Productivity Commission Staff Research Paper, Ausinfo, Canberra.

NMBIPP (2004). *Managing our Natural Resources: Can Markets Help? National Action Plan for Salinity and Water Quality – National Market Based Instruments Pilots Program*. Australian Federal Government. Canberra.

Olmstead, S. (2010). The economics of water quality. *Review of Environmental Economics and Policy*, 4, 44–62.

Pagiola, S. (2008). Payments for environmental services in Costa Rica. *Ecological Economics*, 65, 712–724.

Parkes D., Newell, G. and Cheal, D. (2003). Assessing the quality of native vegetation: the 'habitat hectares' approach. *Ecological Management and Restoration*, 4, S29–S38.

Perrot-Maitre, D. and Davis, P. (2001). *Case Studies of Markets and Innovative Financial Mechanisms for Water Services from Forests*. Forest Trends, Washington, DC.

Productivity Commission (2010). *Market Mechanisms for Recovering Water in the Murray-Darling Basin*. Final Report, March.

Rametsteiner, E. and Simula, M. (2003). Forest certification – an instrument to promote sustainable forest management? *Journal of Environmental Management*, 67, 87–98.

Ribaudo, M. and Nickerson, C. (2009). Agriculture and water quality trading: exploring the possibilities. *Journal of Soil and Water Conservation*, 64, 1–7.

Ribaudo, M., Greene, C., Hansen, L. and Hellerstein, D. (2010). Ecosystem services from agriculture: Steps for expanding markets. *Ecological Economics*, 69, 2085–2092.

Sala, O.E., Chapin III, F.S., Armesto, J.J., et al. (2000). Global biodiversity scenarios for the year 2100. *Science*, 287, 1770.

Salzman, J. (2005). Creating markets for ecosystem services: Notes from the field. *New York University Law Review*, 80, 870–961.

Salzman, J. and Ruhl, J. (2000). Currencies and the commodification of environmental law. *Stanford Law Review*, 53, 607–694.

Schwarze, R. and Zapfel, P. (2000). Sulfur allowance trading and the regional clean air incentives market: a comparative design analysis of two major cap-and-trade permit programs? *Environmental and Resource Economics*, **17**, 279–298.

Srinivasan, U.T., Carey, S. P., Hallstein, E., et al. (2008). The debt of nations and the distribution of ecological impacts from human activities. *Proceedings of the National Academy of Sciences USA*, **105**, 1768–1773.

Stavins, R. (1995). Transaction costs and tradable permits. *Journal of Environmental Economics and Management*, **29**, 133–148.

Stavins, R.N. (1998). What can we learn from the grand policy experiment? Lessons from SO2 Allowance trading. *Journal of Economic Perspectives*, **12**, 69–88.

Stavins, R.N. (2000). *Experience with Market Based Environmental Policy Instruments*. Resources for the Future Discussion Paper 0009, January 2000.

Stoneham, G., Chaudri, V., Ha, A. and Strappazzon, L. (2003). Auctions for conservation contracts; an empirical examination of Victoria's BushTender trial. *Australian Journal of Agricultural and Resource Economics*, **47**, 477–500.

US Environment Protection Agency (EPA) (2001). *The National Costs of the Total Maximum Daily Load Program*. EPA-841-D-01-003. Washington, DC.

US Environment Protection Agency (EPA) (2012). *Compensatory Mitigation Fact Sheet*. Available at: http://water.epa.gov/lawsregs/guidance/wetlands/wetlandsmitigation_index.cfm (accessed August 2012).

Weitzman, M.L. (1974). Prices vs. Quantities. *Review of Economic Studies*, **41**, 477–491.

Whitten, S., Coggan, A., Reeson, A. and Shelton. D. (2007). *Market Based Instruments for Ecosystem Services in a Regional Context*. Report for the Rural Industries Research and Development Corporation (RIRDC) and Land and Water Australia. RIRDC, Canberra.

Whitten, S.M. and Young, M. (2004). Closing plenary session: Market based tools for environmental management: Where do they fit and where to next? In: *Market Based Tools for Environmental Management*. Proceedings of the 6th Annual AARES National Symposium, 2003 (eds S.M. Whitten, M. Carter and G. Stoneham).

Wu, J. (2000). Slippage effects of the conservation reserve program. *American Journal of Agricultural Economics*, **82**, 979–992.

Zedler, J.B. (1996). Ecological issues in wetland mitigation: an introduction to the forum. *Ecological Applications*, **6**, 33–37.

Epilogue: Equitable and Sustainable Systems

Agriculture and urban areas are by far the largest users of ecosystems and their services. Nearly half of the human population depends on agriculture as a source of livelihood. At the same time, the proportion of the human population living in cities has increased from less than 15% at the beginning of the twentieth century to 50% currently. It is estimated that world food demand and the urban population will double by 2050. This will have enormous consequences for already stressed ecosystems and the resources they provide for us. It is important to take action now and incorporate ecosystem thinking into decision-making processes at all scales, from local to global.

The community, through its international agencies, has achieved much in terms of raising awareness and the setting of global targets to bring about a halt in environment degradation since the United Nations Conference on Environment and Development, held in Rio de Janeiro in 1992. Although Agenda 21 has been adopted by members of the United Nations, the programmes to achieve development goals initiated in 2000, through Millennium Development Goals, fall short of their target to reduce poverty by half and ensure food security and environmental integrity, among others. Similarly, the Millennium Ecosystem Assessment raised awareness of ecosystems and their services but the global environment continues to degrade because of a lack of any coherent plan of action. Recently, the United Nations has established the Intergovernmental Science-Policy Platform on Biodiversity and Ecosystem Services to translate ecosystem science into action, and to track the drivers and consequences of ecosystem change world-wide. It aims to do this in consultation with governments and research partners. This

Ecosystem Services in Agricultural and Urban Landscapes, First Edition. Edited by Steve Wratten, Harpinder Sandhu, Ross Cullen and Robert Costanza.
© 2013 John Wiley & Sons, Ltd. Published 2013 by John Wiley & Sons, Ltd.

action plan is focused on strengthening assessment, relevant policy and associated science at spatial and temporal scales.

Apart from understanding the drivers of ecosystem change and the science challenges resulting from it, greater emphasis is required on understanding the social aspects through the lens of equity, justice and sustainability. The struggle for scarce resources may result in conflicts in social–political domains, enlarging the divide between rich and poor and between developed and developing countries. A new paradigm shift at the global level indicates a move towards green economy, as discussed at the Rio + 20 summit. Therefore, the setting up of sustainable development goals under the auspices of the United Nations requires much deeper scrutiny in terms of the consumption of resources, wealth distribution within and between countries and the goal of greater equitability and sustainability.

Managing 6.5 billion people in cities and providing food to 9 billion worldwide by 2050 will need greater coherence in global efforts, partnerships of developed and developing countries, careful planning and implementation of the required programmes with science and policy collaboration.

In this book, we have highlighted the current global challenge to halt ecosystem degradation and provided updated knowledge of two crucial systems – agriculture and urban areas, as well as their incontestable dependence on the 'natural' environment. This book is an integrated attempt by geographically dispersed researchers working in their own disciplines in the hope that it will help to better understand these 'engineered systems' and to improve their management for sustainable human welfare. In the context of a disturbed world, market-dominated economies, little emphasis on human well-being compared with the never-ending pursuit of GDP, and the disaster that one billion people, mainly children, are chronically hungry, are we optimistic? Yes, but very cautiously. What is your contribution to this severe challenge to the future of mankind?

Steve Wratten, Harpinder Sandhu,
Ross Cullen, Robert Costanza
May 2012

Index

Ecosystem Services in Agricultural and Urban Landscapes, First Edition. Edited by Steve Wratten,
Harpinder Sandhu, Ross Cullen and Robert Costanza.
© 2013 John Wiley & Sons, Ltd. Published 2013 by John Wiley & Sons, Ltd.